MEMBRANE STRUCTURE AND FUNCTION

W.H.Evans

National Institute of Medical Research, Mill Hill, London NW7 1AA, UK

J.M.Graham

Department of Biochemistry, St George's Hospital Medical School, London SW17 0RE, UK

IRL PRESS
—at—
OXFORD UNIVERSITY PRESS
Oxford New York Tokyo

IRL Press at Oxford University Press
Walton Street, Oxford OX2 6DP

First published 1989
Reprinted 1989, 1991

British Library Cataloguing in Publication Data

Evans, W. Howard
  Membrane structure and function.
  1. Organisms. Cells. Membranes
  I. Title   II. Graham, J. M.
    574.87'5

ISBN 0 19 963004 6

Library of Congress Cataloging-in-Publication Data

Evans, W. Howard
  Membrane structure and function / W.H. Evans, J.M. Graham.
      p.   cm. — (In focus)
  Includes index.
  1. Cell membranes. I. Graham, J. M. (John M.), 1943 –
II. Title.   III. Series: In focus (Oxford, England)
  [DNLM: 1. Cell Membrane—physiology.
          2. Cell Membrane—ultrastructure.   QH 601 E92m]
QH601.E823   1989   574.87'5—dc20

Previously announced as:

ISBN 1 85221 099 0
ISBN 0-19-963004-6 (pbk.)

Typeset and printed by Information Press Ltd, Oxford, England.

Preface

In this mini-text on membranes, knowledge of the structure of biological membranes has been fused with some of the latest views on how they function. The text aims to illustrate how biology is becoming unified at the membrane level. For example, it is now clear that the actions on cells of hormones, growth factors, neurotransmitters, drugs, toxins and morphogens occur through common membrane-mediated processes. This conservation of biochemical mechanisms becomes increasingly evident as more is learnt about the structure of membrane receptors, ion channels and carriers, as well as the various operational mechanisms in the plasma membrane and elsewhere for amplifying and interpreting biological information.

A text aiming to be up to date often treads that thin line separating fact, conjecture and fiction. The various diagrams presented in this book, depicting the structure in the membrane of receptors, ion pumps and channels, are to be regarded as working models in the process of refinement, and some may even require major modification. Although molecular details are emphasized, we have retained a cell-biological flavour, especially with regard to membrane biogenesis and the topogenic sequences that target proteins to various locations in the cell. We also describe, where appropriate, membrane modifications underlying disease processes, for example receptors governing the uptake by cells of lipoproteins from the blood and the mechanism of action of toxins.

Finally, we would like to thank colleagues for their help in providing information and advice as we strived to concentrate this account into a format that we hope will prove acceptable to young and old students of membrane structure and function.

W.H.Evans
J.M.Graham

Contents

4. Membrane biogenesis and trafficking

5. Membrane transport and bioenergetics

Abbreviations

CAM	cell adhesion molecule
cAMP	cyclic adenosine 5′ monophosphate
CSF	colony stimulating factor
DPG	diphosphatidylglycerol
EGF	epidermal growth factor
$FADH_2$	reduced flavine adenine dinucleotide
GABA	γ aminobutyric acid
Gal	galactose
GalNAc	N-acetylgalactosamine
Glc	glucose
GlcNAc	N-acetylglucosamine
IgA	immunoglobulin A
IGF-1	insulin-like growth factor
IP_3	inositol-1,4,5-trisphosphate
LDL	low-density lipoprotein
MHC	major histocompatibility antigen
NAD	nicotinamide adenine dinucleotide
NANA	N-acetylneuraminic acid
PC	phosphatidylcholine
PDGF	platelet-derived growth factor
PE	phosphatidylethanolamine
PG	phosphatidylglycerol
PI	phosphatidylinositol
PS	phosphatidylserine
Q	ubiquinone (coenzyme Q)
RER	rough endoplasmic reticulum
SDS	sodium dodecyl sulphate
SDS – PAGE	SDS – polyacrylamide gel electrophoresis
SER	smooth endoplasmic reticulum
SM	sphingomyelin

1

General features and properties

1. Introduction

Membranes are essential components of all cells. While eukaryotes may contain a wider range of membranous structures than do prokaryotes, their general function and structure is universal. Cell membranes have four main functions which are inter-related.

(i) They form the physical boundaries of compartments whose composition can be controlled to permit biochemical processes to occur efficiently.

(ii) Their structure allows the transport of a restricted range of molecules from one compartment to another, that is they are selectively permeable.

(iii) They are interfaces which transduce chemical signals or energy from one compartment to another.

(iv) They provide the optimum environment for the functioning of molecules (e.g. enzymes, ion pumps and receptors) which are associated with the functions in (ii) and (iii).

Membranes are distinguished by their amphipathic nature. All membranes in the cell are based on a general structure which is hydrophobic (non-polar) in the middle and hydrophilic (polar) on the outside. The molecules which make up this structure vary with the type of organism and, in eukaryotes, there are also variations between the different types of membrane within the cell. The molecular composition and architecture of these membranes will be described in Chapter 2. Suffice it to state here that membranes are approximately 5 nm in thickness and they contain protein and lipid molecules. Many of the proteins and lipids have carbohydrate residues covalently attached to them; these are called glycoproteins and glycolipids.

2. Surface (plasma) membrane

All cells are surrounded by a surface membrane (*Figure 1.1*), usually called the plasma membrane in animal cells. In other organisms, terms such as cytoplasmic membrane, limiting membrane, protoplast, plasmalemma or cell envelope are used depending on the type of cell. This membrane provides the reactive interface

1

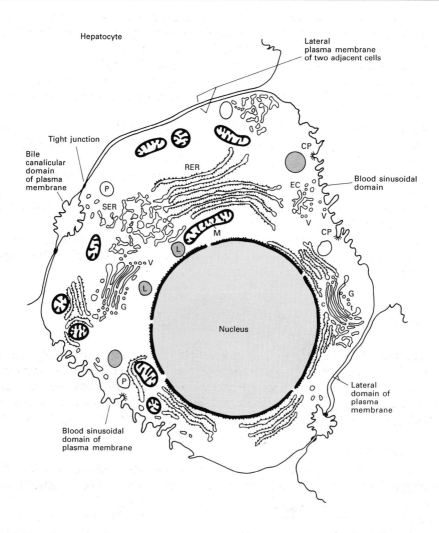

Figure 1.1. Section through hepatocyte illustrating the various organelles and membrane systems. EC, endocytic compartment; G, Golgi apparatus (c = *cis* face; t = *trans* face); L, lysosome; M, mitochondrion; P, peroxisome; RER, rough endoplasmic reticulum; SER, smooth endoplasmic reticulum; V, vesicle; CP, coated pit.

between the external medium and the cytoplasm. It is the membrane across which all nutrients must pass into the cell and across which all molecules to be excreted and secreted must pass. It is the membrane which provides the initial response of the cell to all extracellular stimuli. The plasma membrane of animal cells is freely permeable to a restricted range of small non-ionized molecules such as water, urea, oxygen, carbon dioxide and glycerol and to many lipid-soluble molecules. However, specific transport systems are needed for molecules such as glucose and amino acids. Enzymes which are responsible for transporting specific ions are also present in surface membranes. The Na^+/K^+-ATPase, for example, uses the energy of ATP hydrolysis to pump Na^+ out of the cell in exchange for K^+; likewise the Ca^{2+}-ATPase enzyme pumps Ca^{2+} out of the cell. Peptide hormones such as insulin and glucagon bind to the external surface and elicit a response from the internal surface by virtue of the presence in the plasma membrane of proteins to which the hormones bind and transduce the hormonal signal into the cell.

The surface membrane of epithelial cells is divided into 'domains' or regions, which are morphologically distinct as shown in *Figure 1.1*. The plasma membrane of epithelial cells (e.g. in the gut and kidney) is segregated by a tight junction into apical and basolateral domains. In the hepatocyte, the basolateral membrane is further divided into a blood-bathed, sinusoidal domain and a lateral domain interacting with neighbouring cells; the apical plasma membrane, enclosing the bile spaces, is called the bile canalicular membrane.

Prokaryotes rarely have any membrane structures other than the surface or limiting membrane. In bacteria the system which provides the cell with energy is also closely associated with this membrane.

3. Internal membranes

Unlike prokaryotes, eukaryotes have a variety of membrane-bound organelles within their cytoplasm. The membranous organelles of animal cells are described in this section. These structures are present in virtually all types of eukaryotic cells—indeed, plant cells contain an even larger variety of cytoplasmic bodies including tonoplasts and plastids, among which chloroplasts are the most notable.

3.1 Mitochondria

In eukaryotes the energy-transduction system is associated with the mitochondria (in plants this function is also associated with chloroplasts). The mitochondrion is a double membrane structure ($\sim 1\ \mu m$ in diameter). The central matrix compartment contains all the enzymes associated with final oxidation of carbohydrate, lipid and protein metabolites, while the inner mitochondrial membrane, which is highly folded into cristae, contains the electron transport system. This is linked to the movement of H^+ from the matrix into the intermembranous space and an ATP synthase enzyme is linked to the movement of H^+ back into the matrix. The membranes also contain transport systems for moving carbohydrate, protein breakdown products and acyl chains from cytoplasm to matrix and for exchanging ATP and ADP; a tricarboxylic acid transporter moves acetate (as citrate) out of the matrix.

Genetic defects in the electron transport system of the inner mitochondrial membrane are now recognized as a cause of myopathy. The clinical features of mitochondrial myopathies include muscle weakness, dementia, ataxia, movement disorders and peripheral neuropathy. These mitochondrial myopathies usually involve modifications to either the NADH-Q reductase or the ubiquinol–cytochrome c reductase, although defects in both cytochrome c oxidase and ATP synthase have been reported. Another important clinical aspect of mitochondrial membrane function is associated with the mode of action of anaesthetics. Halothane, for example, a widely used general anaesthetic, acts on the inner mitochondrial membrane, causing amongst other effects the uncoupling of phosphorylation from oxidation.

3.2 Lysosomes

Lysosomes contain a wide variety of hydrolytic enzymes (e.g. phosphatases, sulphatases, esterases, proteases and glycosidases) that digest polypeptides, polysaccharides, polynucleotides and glycolipids, yielding monomer units. Lysosomes are surrounded by a membrane that prevents the enzymes from diffusing out and digesting other cell components.

Lysosomes comprise the major intracellular digestive system, and elaborate mechanisms exist for the transfer of macromolecules into the lysosome for hydrolysis. The membranes of the endocytic compartment are believed to play a central role in controlling entry into lysosomes (see Section 5.2 of Chapter 4).

One common feature of the enzymes in lysosomes is that they have a pH optimum of 5–6 generated and maintained by a proton-translocating ATPase. A low intraluminal pH is also found in the endocytic compartment and in secretory vesicles where it is believed that a similar enzyme is responsible for maintaining a low pH.

A range of inherited disorders (collectively called lysosomal storage diseases) have been discovered in which the absence of one or more of the hydrolytic enzymes leads to the accumulation in lysosomes of macromolecules and consequent disruption of cell function. One example is Tay–Sach's disease in which there is a deficiency of β-hexosaminidase, leading to an accumulation of gangliosides inside the cell. Other deficiencies in lysosomal function relate to the malfunctioning of enzymes in the lysosomal membrane that transport sugars, amino acids, etc., across it.

In addition to the uptake and degradation of macromolecules, lysosomes also engage in autophagy, involving the sequestration of intracellular material. Morphological studies show that lysosomes invaginate around other membrane-bound organelles and then fuse with the membranes. Finally, the organelles disintegrate as they are hydrolysed inside these autophagic vacuoles.

3.3 Peroxisomes

Peroxisomes (microbodies or glyoxysomes) are surrounded by a single membrane. They contain a variety of oxidases which act upon a broad range of substrates including D-amino acids and, in the kidney, L-amino acids and hydroxyl acids such as glycolate and lactate. The common feature of all these oxidases is that they produce hydrogen peroxide from oxygen. Catalase, which constitutes 15% of the total peroxisomal protein, reduces the hydrogen peroxide to water. Catalase, while it is used as a characteristic marker for peroxisomes, is also present in the cytoplasm of many cells and is also involved in the final

step of detoxification of oxygen radicals. Free radicals are potentially damaging
to the cell and to neutralize these the enzyme superoxide dismutase produces
hydrogen peroxide from protons and oxygen radicals. The hydrogen peroxide
is then detoxified to water by catalase.

Peroxisomes share some properties with mitochondria. For example, they carry
out a cyclical series of reactions which use and regenerate isocitrate. In the
glyoxylate cycle present in plants, isocitrate is cleaved to yield succinate and
glyoxylate; malate is generated from the condensation of glyoxylate and acetyl
CoA and then isocitrate is regenerated from malate in the same manner as in
the Krebs cycle.

Peroxisomes also carry out β-oxidation of fatty acids. The series of metabolic
intermediates produced is identical to those produced by β-oxidation in
mitochondria but the enzymes are quite distinct. The reduced flavine adenine
dinucleotide (FADH$_2$) produced in the first oxidative step is re-oxidized by
oxygen with the production of hydrogen peroxide.

Peroxisomes are also involved in the synthesis of certain lipids. They contain
dihydroxyacetone phosphate acyl transferase and alkyl dihydroxyacetone
phosphate synthase which are important enzymes in the synthesis of ether lipids
(plasmalogens). These are phospholipids in which one of the hydrocarbon chains
is linked to the glycerol backbone through an ether link rather than an ester
line (see Section 1.3 of Chapter 2).

3.4 Nuclear membranes

The genetic material of eukaryotes is surrounded by a nuclear membrane (except
during cell division). It is a double membrane with large, well-defined pores.
Figure 1.1 shows that there is continuity between the outer and inner membranes
in contrast to the outer and inner membranes of mitochondria which are spatially
discrete. The pore appears to regulate the movement of macromolecules between
the nuclear matrix and the cytoplasm. Movement of proteins into the nucleus
is an ATP-dependent process and also requires the presence of specific
addressing amino acid sequences (see Section 4.4 of Chapter 4). The inner nuclear
membrane contains attached chromatin, whereas the outer membrane, studded
with ribosomes, appears, in places, to show continuity with the endoplasmic
reticulum. Nuclear pores have been studied in frog oocytes using high resolution
electron microscopy. As shown in Figure 1.2, the pore with a diameter of 7 nm
contains a central plug and is surrounded by eight symmetrical subunits arranged
at the cytoplasmic end. Microinjection experiments using dextrans, colloidal gold
and various proteins indicate that the nuclear pore allows passive movement
of molecules up to 9 nm diameter, indicating its flexible nature.

3.5 Endoplasmic reticulum and Golgi membranes

The membranes of the endoplasmic reticulum and Golgi separate the cytoplasm
from the cisternal space. The rough endoplasmic reticulum, which is probably
continuous with the nuclear membrane, bears ribosomes on its cytoplasmic face
(Figure 1.1) and is the site of membrane protein synthesis. The smooth

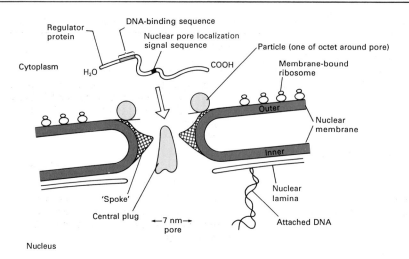

Figure 1.2. Model of the nuclear pore. A DNA regulator protein (e.g. steroid hormone receptor) is about to bind to the pore prior to translocation into the nucleus.

endoplasmic reticulum synthesizes membrane lipids (cholesterol and phospho-lipids). The sequential addition of carbohydrate residues to proteins and lipids occurs in both the smooth endoplasmic reticulum and the Golgi membranes.

The core sugars are added to glycoproteins in the smooth endoplasmic reticulum. The membranes comprising the Golgi apparatus contain glycosyl and sialyl transferases which add the terminal sugars to glycoproteins and glycolipids. These enzymes are associated predominantly with the *trans*-Golgi (that part of the Golgi stack nearest the plasma membrane) as distinct from the *cis*-Golgi (that part of the Golgi stack nearest the nucleus).

The Golgi apparatus is also important in assembling lipids and proteins into the correct conformation for incorporation into membranes or for secretion. It is also involved in the processing of pro-hormones involving a reduction in their size, for example the conversion of pro-insulin to insulin.

The Golgi apparatus features in the synthesis of proteoglycans—large, extensively glycosylated polypeptides which are important in the make-up of the cell coat or glycocalyx. The terminal reactions involved in the production of proteoglycans result in the modification of their long carbohydrate chains. These include epimerization, de-acetylation and sulphation. The enzymes for these reactions also occur predominantly in the *trans*-Golgi membranes.

3.6 Membrane vesicles

Extensive vesicular traffic occurs inside cells. These membrane vesicles transport proteins and lipids to specific destinations in the cell, and act as carriers in exocytosis and endocytosis. In exocytosis, vesicles are observed to move from the Golgi apparatus to the cell surface; in secretory cells, these 'secretory' vesicles are often filled with contents, for example lipoproteins, which are released to the outside (see Section 3 of Chapter 4). In a reverse process, endocytosis, the

Table 1.1. Relative amounts of different membrane types in the rat hepatocyte

Subcellular membrane/organelle	% of total membrane[a]	Comments
Plasma membrane	5	Polarized to form canalicular (bile facing), lateral (cell facing), and sinusoidal (blood facing) domains measuring ~10%, 40% and 50% of the total surface area, respectively.
Endoplasmic reticulum		
rough	30	Can vary according to metabolic state.
smooth	14	Induced to proliferate by drugs.
Golgi apparatus	6	Rough estimates; distinction between
Endocytic compartment (vesicles)	3	*trans* compartment of Golgi apparatus, endocytic compartment and pinocytic vacuoles is unclear.
Lysosomes	1	
Peroxisomes	1	
Mitochondria		
outer membrane	7	
inner membrane	30	
Nuclei (inner membrane)	0.3	Outer membrane shows continuity with rough endoplasmic reticulum

[a]The above values are estimates based on the amount of protein recovered in different subcellular fractions from liver homogenates and upon morphometric quantification of observations in light and electron microscopes. Note that a hepatocyte can contain ~1000–2000 mitochondria, a few hundred lysosomes and peroxisomes, several million ribosomes of which ~75% are bound to membranes. Like the nucleus, the Golgi apparatus is a single organelle. About 150 acidic (mainly endocytic) compartments per hepatocyte are identified by fluorescence microscopy.

vesicles (endosomes) pinch off from the plasma membrane and migrate into the endocytic compartment. This is an area of cell consisting of ramifying tubular processes and vesicles, and is a major site of control of vesicular trafficking (see Section 5 of Chapter 4). Vesicles covered by a coat of clathrin are also observed, especially near the plasma membrane, where they are formed from coated pits, and in the Golgi apparatus environs where they are smaller in diameter and are involved in exocytosis.

In protozoa, large vesicles (vacuoles) are seen which are specialized for the ingestion of nutrients and storage of water and waste products. In macrophages, large intracellular vesicles are involved in the ingestion of bacteria, solid material, etc. (phagocytosis).

The approximate amount of each type of membrane found in a typical animal cell is shown in *Table 1.1*.

4. Purification and characterization of membranes

Important contributions to our knowledge of the structure and function of

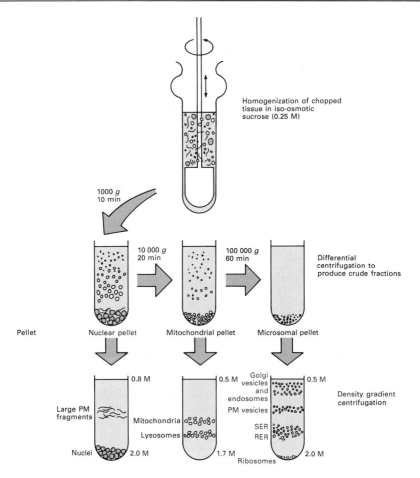

Figure 1.3. Generalized scheme showing subcellular fractionation of a mammalian tissue, e.g. liver. Figures are molar sucrose concentrations at top and bottom of gradients. PM, plasma membranes; SER, smooth endoplasmic reticulum; RER, rough endoplasmic reticulum.

membranes have been provided by the ability to isolate highly purified membranes and organelles. For example, the isolation of mitochondria and chloroplasts in high purity and appreciable yield has allowed their complex membrane structure and the underlying bioenergetic mechanisms to be dissected. Similarly, the isolation of plasma membranes and various endo-membranes as closed vesicles has allowed the mechanism of membrane transport and receptor-mediated transmembrane signalling to be studied under precise experimental conditions.

In simplest terms, the isolation of subcellular membranes entails the following consecutive processes (*Figure 1.3*).

(i)   Cell breakage. This provides, ideally, intact organelles (nuclei, lysosomes,

Table 1.2. Subcellular markers for animal cell membranes e.g. hepatocytes

Membrane	Enzymatic marker	Chemical marker	Comments
Plasma membrane	Adenylate cyclase (hormone-activated), Na^+/K^+·ATPase	High cholesterol and sialic acid	Many receptors and antigens are present
Golgi apparatus	Glycosyl transferases Thiamine pyrophosphatase		Most markers apply to *trans* aspect
Endocytic compartment	None identified	–	High amounts of receptors and endocytosed ligands
Lysosomes	Many hydrolases, e.g. acid phosphatase, aryl sulphatase	–	–
Endoplasmic reticulum	Glucose-6-phosphatase	Cytochrome b_5 P450	Markers vary between tissues
Mitochondrial membranes	Outer—monoamine oxidase Inner—cytochrome oxidase	Cardiolipin	
Nuclear membrane	None identified	Histone RNA	Complex double membrane with pores
Peroxisomes	Catalase		

mitochondria) and vesicles derived from the endoplasmic reticulum, Golgi and endocytic apparatus. Plasma membrane may form large sheet-like fragments or vesicles.

(ii)   Separation. The separation of the various components dispersed in the cell or tissue homogenate is usually achieved by application of centrifugal forces (*Figure 1.3*). These methods exploit differences in the size and mass of the membrane components and are usually followed by further centrifugation in gradients constructed of inert solutions to separate membranes according to their density. These density differences are a function, mainly, of the ratio of protein to lipid in the various membranes. Sucrose solutions are usually used to prepare these density gradients, but other, special types of gradient materials are also used, for example, Percoll, consisting of silica particles covered in polyvinyl pyrrolidone and derivatives of tri-iodobenzoic acid such as metrizamide and Nycodenz. Use of these newer types of gradient media can often give more rapid separations than sucrose gradients.

Other procedures used to separate membranes include machines generating electrophoretic fields that exploit membrane surface charge differences and the use of antibodies to separate membranes according to their affinity towards their respective antigens. One can distinguish traditional methods that exploit physical differences from affinity methods that exploit the biological properties of membranes.

Identification of membranes is possible because various membranes and the organelles possess specific marker components—mainly enzymatic (see *Table 1.2*). In animal cells, approximately 10 types of biochemically discrete membranes exist, whereas in plant cells up to 13 types of membrane have been described. Prokaryotic cells are correspondingly simpler, with only two major types of membrane identified in bacteria.

5. Further reading

Microbial membranes

Rogers,H.J., Perkins,H.R. and Ward,J.B. (1980) *Microbial Cell Walls and Membranes.* Chapman & Hall, London.

Plant organelles

Harwood,J.L. and Walton,T.J. (eds) (1988) *Plant Membranes—Structure, Assembly and Function.* Biochemical Society, London.
Quinn,P.J. and Williams,W.P. (1983) *Biochim. Biophys. Acta,* **737**, 223–266.
Murphy,D.J. (1986) *Biochim. Biophys. Acta,* **864**, 33–94.

Membrane isolation

Graham,J.M. (1984) In *Centrifugation: A Practical Approach.* Rickwood,D. (ed.), IRL Press, Oxford, p. 161–182.
Biological Membranes: A Practical Approach (1987) Findlay,J.B.C. and Evans,W.H. (eds), IRL Press, Oxford.

2

Composition and structure

1. Lipids

1.1 General composition

Membranes are constructed of a bilayer of amphipathic lipid made up of phospholipids and glycolipids (*Figure 2.1*). The relative amounts of these two lipids varies with the type of membrane (*Table 2.1*). Almost without exception in prokaryotic membranes, phospholipid is the only major lipid component, while eukaryotic membranes may also contain a sterol; in plants this is usually ergosterol while in animals one finds cholesterol (*Figure 2.1*). If cholesterol is included in the growth medium of prokaryotes, it will be taken into the protoplast membrane, but only some mycoplasmas have a requirement for cholesterol. Plant chloroplast membranes also contain significant amounts of sulpholipids. The sterol in membranes is unesterified in contrast to the esterified form found in serum lipoproteins. The nuclear membrane of animal cells may contain low amounts of cholesterol ester. The relative amounts of cholesterol and phospholipid in animal cell membranes varies widely (*Table 2.1*), being highest in the plasma membrane and lowest in the intracellular membranes. In human fibroblast membranes approximately 95% of cholesterol resides in the plasma membrane.

1.2 Structure of amphipathic lipids

The major classes of membrane amphipathic molecules of prokaryotes and eukaryotes are based predominantly on diacylglycerol. In animal cells they are also based on ceramide (*N*-acylsphingosine). In phospholipids the free primary hydroxyl group of the glycerol (or sphingosine residue) is phosphorylated. The molecular forms of phospholipid (*Figure 2.2*) reflect the variety of residues that can be linked to this terminal phosphate group. Bacteria have a huge range of such compounds which are determined largely by the culture conditions but the dominant forms are phosphatidylglycerol, diphosphatidylglycerol and phosphatidylethanolamine. In plants, phosphatidylglycerol predominates along with smaller amounts of others such as phosphatidylcholine, -ethanolamine and -inositol. In animal cells the phospholipid composition varies according to membrane type. For example, sphingomyelin is a major phospholipid in the plasma membrane while the inner mitochondrial membrane contains

11

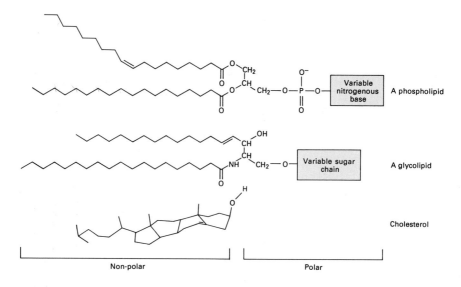

Figure 2.1.   General structure of amphipathic lipids.

Table 2.1.   Phospholipid, glycolipid and cholesterol content of membranes

Membrane	% of total lipid by weight		
	Glycolipid	Phospholipid	Cholesterol
Human erythrocyte	11	61	22
Myelin	28	41	22
Rat liver mitochondria	<5	80	4
Endoplasmic reticulum	<5	75	8
Bacterial protoplast	trace	80–90	0
Plant chloroplast	80	12	0

diphosphatidylglycerol (cardiolipin) (see *Figure 2.3*).

Other amphipathic lipids based on either ceramide or diacylglycerol (the glycolipids) display a variety of forms (*Figure 2.4*). The free primary hydroxyl group of either the diacylglycerol or ceramide residue is linked to one or more sugar residues. In plants, the relatively simple monogalactosyl and digalactosyldiglycerides are common. In bacteria the protoplast membrane is largely devoid of glycolipids, although the outer membrane of Gram-negative bacteria, which is peripheral to the protoplast (inner membrane), does contain some very complex glycolipids called lipopolysaccharides.

Animal cells contain a large array of glycolipid molecules, all based on the ceramide structure. The carbohydrate chains may be linear or branched and in addition to galactose and glucose they may contain N-acetylated sugars and sialic acid (N-acetylneuraminic acid; *Figure 2.4*). Cerebrosides contain uncharged sugar residues only; gangliosides also contain N-acetylated sugars and

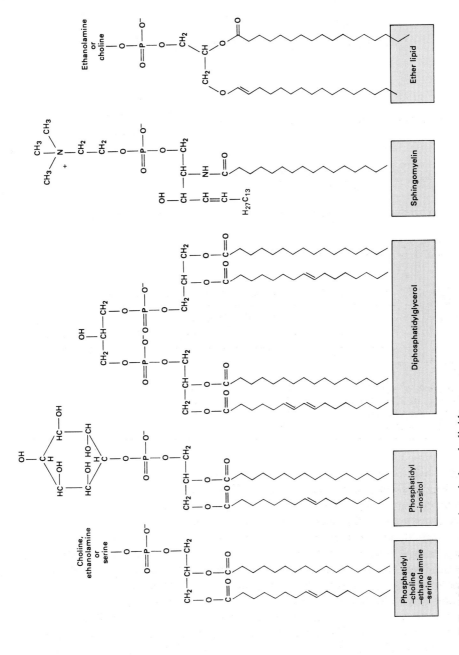

Figure 2.2. Molecular species of phospholipids.

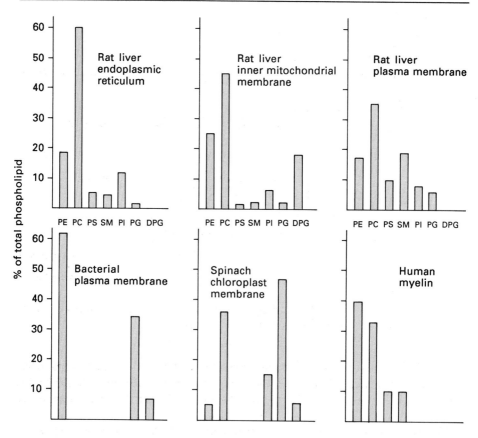

Figure 2.3.   Phospholipid composition of membranes. PE, phosphatidylethanolamine; PC, phosphatidylcholine; PS, phosphatidylserine; SM, sphingomyelin; PI, phosphatidyl-inositol; PG, phosphatidylglycerol; DPG, diphosphatidylglycerol (cardiolipin).

N-acetylneuraminic (sialic) acid residues. All of these amphipathic lipids (other than cholesterol or ergosterol) contain two long hydrocarbon chains. In the ceramide-based lipids, one of these chains is the invariant 15 carbon (C-15) hydrocarbon chain of the sphingosine molecule. However the two acyl chains which are esterified onto glycerol-3-phosphate in glycerol-based lipids and the single amide-linked acyl chain in ceramide-based lipids are variable. The fatty acids range from C-12 to C-24, with one usually saturated and the other unsaturated: the number of carbon:carbon double bonds may be one, two or four. One of the most important fatty acid residues is arachidonate which has four double bonds (C-20:4).

1.3 Ether lipids and dolichol

Ether lipids or plasmalogens are another important class of phospholipids. In these molecules the non-polar hydrocarbon chain at position 1 of glycerol is

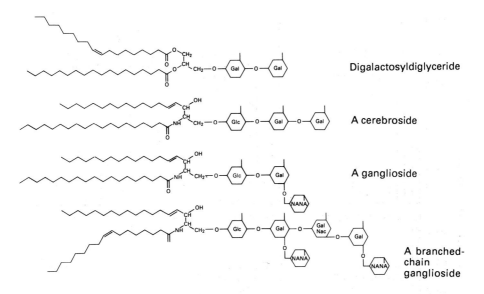

Figure 2.4. Molecular species of glycolipids. Gal, galactose; Glc, glucose; NANA, *N*-acetylneuraminic acid; GalNAc, *N*-acetylgalactosamine.

attached via an ether link rather than an ester link. The ether-linked hydrocarbon chain frequently has a double bond at the 1′ position (see *Figure 2.2*). Both alkyl and alkenyl derivatives of phosphatidylcholine and phosphatidylethanolamine exist.

Although in most membranes ether lipids represent only a very small percentage of the total phospholipid, they are nevertheless major components in certain membranes. For example, in canine myocardial sarcoplasmic reticulum, 53% of the phospholipid is ether lipid and in skeletal sarcoplasmic reticulum over 70% of the ethanolamine-containing phospholipid is ether lipid. Macrophages and platelets are also rich in these molecules whose major fatty acid is arachidonate.

Dolichol phosphate is an important intermediary in the synthesis of glycoconjugates (see Section 2.2 of Chapter 4). Dolichol is a polyisoprenoid compound with the general formula $CH_3 - C(CH_3) = CH - CH_2 - [CH_2 - C(CH_3)$ $= CH - CH_2]_n - CH_2 - CH(CH_3) - CH_2 - CH_2 - R$ where R is a hydroxyl group, monophosphate, pyrophosphate or a fatty acid, and $n = 15 - 19$. This, the largest known lipid, is confined to the membranes of the smooth endoplasmic reticulum where the phosphate group is orientated towards the cisternal space.

1.4 Lipid organization

Most of our detailed knowledge on the organization of amphipathic lipids in the bilayer structure of membranes has come from work on mammalian cell membranes and the human erythrocyte membrane in particular. The non-polar central part of all membranes contains the fatty-acyl chains of phospholipids and

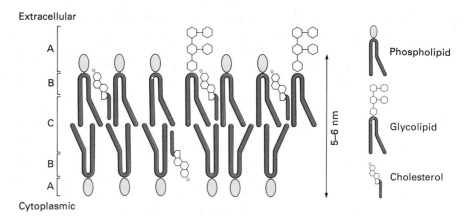

Figure 2.5.   Arrangement of membrane lipids in the bilayer of a plasma membrane. A, polar regions; B, non-polar region—fluidity modulated by cholesterol and alkyl chain unsaturation; C, fluid non-polar region.

glycolipids, whereas the polar phosphodiester-containing residues or the glycosyl chains are on the outside of the membrane. The free hydroxyl group of unesterified sterols is also orientated towards the periphery, while the planar ring system and the aliphatic side chain are orientated towards the centre of the bilayer (*Figure 2.5*). In animal plasma membranes phospholipids and glycolipids are asymmetrically distributed between the two halves of the bilayer. Sphingomyelin and phosphatidylcholine are predominantly in the outer half with the other phospholipids located mainly in the inner half of the bilayer. Phosphatidylinositol is also predominantly positioned in the inner half of the bilayer—this lipid is involved in intracellular signalling (see Section 2.3 of Chapter 3). In *Figure 2.5* cholesterol is shown to be enriched in the outer half of the bilayer, but there is some controversy about its precise distribution. Glycolipids reside exclusively in the outer half; indeed, generally all the glycosyl residues of all carbohydrate-containing components are orientated on the outer aspect of the plasma membrane (see *Figures 2.5* and *2.8*). Intracellular membranes also contain an asymmetrical distribution of phospholipids in the bilayers. The asymmetry of phospholipids in the surface and internal membranes is probably a consequence of the mechanism by which these molecules are synthesized (see Section 2.3 of Chapter 4).

The bulk of the hydrocarbon chains of the lipids are in a fluid state; that is, there is considerable flexing and rotation of the C–C bonds. The degree of fluidity increases with unsaturation of the fatty acid acyl chains and decreases according to the amount of sterol in the membrane. The more unsaturated the acyl chain, the lower the transition temperature between crystalline and fluid states. Cholesterol restricts the mobility of the hydrocarbon chains close to the polar head groups and also inhibits crystallization of the acyl chains below the transition temperature.

Under physiological conditions (35–37°C in mammals) membrane lipids are in a fluid state allowing many lipids and proteins to diffuse laterally in the bilayer.

This fluidity of the membrane is essential for normal functions to occur—for membrane biogenesis, trafficking, exocytosis and endocytosis described in Chapter 4. Selective uptake of materials by the cell and the extensive vesicular traffic occurring inside cells are merely two examples of membrane-mediated processes governed by the fluid state of membranes. For example, uptake of materials into cells diminishes below 20°C and ceases at 4°C. Poikilothermic organisms adapt to low temperatures by modifying their lipid components.

Although lipid molecules in the bilayer are free to move laterally through the plane of the membrane, they rarely flip from one half of the bilayer to the other. A single phospholipid molecule can move about 2 μm in 1 sec. However, 'flip-flop' (the transbilayer movement of a phospholipid molecule), expressed as a half time, varies from 8 h to 1 – 2 days. Since the phospholipid distribution across the bilayer is asymmetric, extensive random 'flip-flop' clearly does not occur.

2. Proteins and glycoproteins

2.1 General composition

All biological membranes contain proteins. The ratio by weight of protein to lipid varies from 3.6 in the inner mitochondrial membrane to 0.25 in the lipid-rich myelin membrane. For most animal cell plasma membranes the ratio is close to 1.0, while the plasmalemma of prokaryotes and the membranes of the plant chloroplast are similar to the mammalian inner mitochondrial membrane. Many of the proteins contain carbohydrate side chains. In mammalian cells the glycoside chain of glycoproteins is commonly attached to the amide group of an asparagine residue. The chain consists of core and terminal sugars. In the synthesis of membrane glycoproteins core sugars are added to the polypeptide chain by enzymes in the smooth endoplasmic reticulum whereas the terminal sugars are added later by enzymes in the Golgi apparatus. The core sugars usually contain the sequence N-acetylglucosamine – N-acetylglucosamine – mannose – mannose whereas the terminal sugars include galactose, N-acetylated derivatives of glucosamine and galactosamine and sialic acid (N-acetylneuraminic acid). Sialic acid is normally the terminal residue and the carbohydrate chains are frequently branched.

The complement of proteins and glycoproteins in a membrane can be analysed by sodium dodecyl sulphate – polyacrylamide gel electrophoresis (SDS – PAGE). SDS is a detergent which solubilizes and confers uniform charge on membrane proteins. Milder detergents such as deoxycholate and Triton are used to solubilize proteins selectively from membranes. In one-dimensional SDS – PAGE, membrane proteins are resolved mainly on the basis of their molecular weights (although glycosylation can retard the electrophoretic migration of some proteins). The polypeptides of the human erythrocyte membrane, analysed by this method, are shown in *Figure 2.6a*. Traditionally the proteins of this membrane are identified numerically. Greater resolution can be obtained by two-dimensional electrophoresis in which the proteins are resolved in the first dimension on the

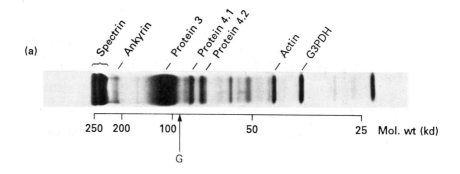

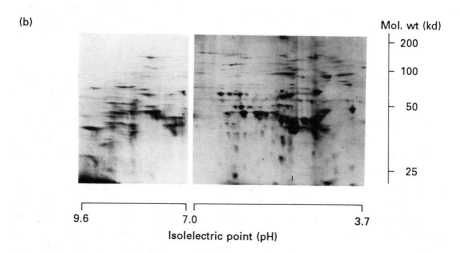

Figure 2.6.   (a) One-dimensional SDS–PAGE electrophoretic separation of human erythrocyte membrane polypeptides (Coomassie Blue stained). (b) Two-dimensional SDS–PAGE electrophoretic separation of rat liver plasma membrane polypeptides (Coomassie Blue stained). G3PDH, glyceraldehyde-3-phosphate dehydrogenase. G marks the position of glycophorin (30 kd). Its high sugar content retards its electrophoretic migration and inhibits staining. (Figure a, courtesy of Dr M.J.A.Tanner.)

basis of their iso-electric points prior to separation in the second dimension by SDS–PAGE. This technique has resolved as many as 70 proteins from rat-liver plasma membranes (*Figure 2.6b*) and because of this complexity proteins are now designated by molecular weight and isoelectric point.

2.2 Protein organization

2.2.1 Disposition in the lipid bilayer

Five categories of membrane proteins exist as shown in *Figure 2.7*. Class I proteins are peripheral or extrinsic proteins such as the F_1 subunit of the ATP-synthase and cytochrome *c*, which lack any well-defined hydrophobic

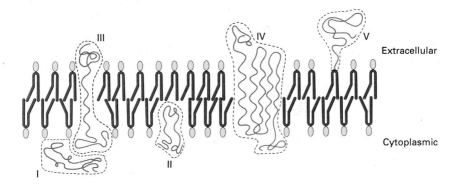

Figure 2.7.   Classes of membrane proteins. I, peripheral protein; II, protein partially inserted into lipid bilayer; III, integral protein with one transmembrane domain; IV, integral protein with five transmembrane domains; V, lipid-anchored peripheral protein.

sequences to anchor them in the lipid bilayer. Peripheral proteins are bound mainly by ionic forces to the polar head groups of phospholipids or to other proteins. Class I proteins can be removed by raising the ionic strength of the medium in which the membranes are suspended. Extracellular matrix and cytoskeletal components associated with the external and cytoplasmic membrane faces, respectively, may be disposed like Class I proteins or they may bind to Class I proteins.

Class II proteins are those which are anchored into a part of the lipid bilayer by a hydrophobic peptide. Evidence for Class II proteins in the outer half of the plasma membrane is lacking. However, toxins which interfere with membrane functions (e.g. mellitin, the bee venom toxin) may insert partially into the extracellular face of the membrane. Also, subunits of the nucleotide-binding G-protein family (see Section 2.5 of Chapter 3) may belong to this class.

Class III and IV proteins are integral (transmembrane) proteins in which largely polar amino acid sequences located at the external and cytoplasmic aspects of the membrane interact with phospholipid head groups and the extracellular and intracellular environments. These domains are connected by a single (Class III) or several (Class IV) hydrophobic peptide domains (of 20 – 30 amino acids) which span the lipid bilayer. The erythrocyte membrane protein glycophorin is the archetypal example of a Class III protein. Many receptors belong to this category. Some Class IV proteins function as transmembrane channels and ion pumps. Because transmembrane proteins have a defined orientation in the membrane, they display a functional asymmetry. For example, the K^+- and ouabain-binding sites of the Na^+/K^+-ATPase are located on the extracellular surface, while the Na^+-binding and ATP-hydrolysing sites are on the cytoplasmic surface. These proteins can only be solubilized by extraction of membranes with 8 M urea or 6 M guanidinium chloride or with detergents (e.g. SDS).

A relatively new class of membrane protein (Class V) in which a peripheral protein domain is anchored to the lipid bilayer by a covalently attached glycolipid has now been recognized. Certain plasma membrane receptors of protozoa and mammalian cells fall into this category and these are described in Section 2.2.4. A further sub-class features anchoring by a fatty acid attached to the protein by a thio-ester bond. For example, palmitic acid is attached to a cysteine residue as found in the transferrin receptor and the *ras* oncogene product. Myristic acid is attached to the amino terminus of the *sarc* oncogene product and the α subunit of the nucleotide-binding G-protein.

2.2.2 Determining the topography of membrane proteins

The topography of proteins and glycoproteins within the membrane, particularly that of transmembrane proteins, imposes some strictures on the amino acid sequences of their peptide chains. The prediction of the two-dimensional arrangement of the linear amino acid sequence in the membrane lipid bilayer entails the application of a number of complementary approaches.

(i)   The linear amino acid sequence of a protein is deduced either directly by chemical methods or indirectly from the sequence of its mRNA. Usually membrane proteins can be isolated in small quantities and it is often necessary to use gene cloning methods to obtain the complete sequence. However, sequences of 50 amino acids or more at the amino terminus of the protein have been obtained by classical Edman sequential degradation techniques using automated gas-phase sequencers. The limited sequence information derived by these chemical methods is then used to generate anti-peptide antibodies or to synthesize deoxyribonucleotide probes chemically both of which can be used to screen cDNA libraries and identify the appropriate clones. The presence in the linear amino acid sequence of an extended series of hydrophobic amino acids is suggestive of a transmembrane segment. Computer programs permit the construction of hydropathy plots which deduce the relative hydrophobicity along the amino acid sequence so that the positions of transmembrane domains and extracellular and cytoplasmic loops can be predicted.

(ii)   Aiding in the deduction of the arrangement of proteins in membranes is the knowledge that amino acid sequences bearing carbohydrate side chains are invariably located at the extracellular face of the plasma membrane (and at the luminal face of intracellular membranes). Consensus glycosylation sites (see Section 2.2 of Chapter 4) provide clues as to where carbohydrate side chains may be attached. Membrane proteins are usually phosphorylated at the cytoplasmic face at specific serine, threonine or tyrosine residues.

(iii)   Chemical or antibody reagents are used to label and identify sequences exposed at either the extracellular or cytoplasmic face of the lipid bilayer, in cells and isolated membrane vesicles. Antibodies raised to peptides corresponding to specific amino acid sequences in the protein are used to locate the exact position of the sequence on membranes (i.e. whether it is intracellular or extracellular).

(iv)   A further approach entails the use of specific proteases to remove peptides comprising external or cytoplasmic loop sequences. The polypeptide sequences remaining within the membrane can then be analysed.

(v)   To deduce the crucial aspects of the sequence in relation to the functioning of proteins, protein engineering techniques such as site-directed mutagenesis to change or delete specific amino acids or sections of the sequence are being applied. For example, the function of the various domains of low-density lipoprotein receptor has been clarified by use of humans in whom natural genetic modifications to the receptor structure have occurred (see Chapter 3, Section 3). However, this approach is still in its infancy because there is insufficient information on the three-dimensional organization of most proteins at present.

The next stage in studying the topography of membrane proteins requires the deduction of their three-dimensional arrangement. Since membrane proteins crystallize only with difficulty, precise knowledge of their overall organization is limited at present to a few examples (e.g. bacteriorhodopsin).

2.2.3 Topography of Class III and IV proteins

Transmembrane proteins are selectively enriched in non-polar amino acids such as valine, isoleucine, leucine, tryptophan and phenylalanine in one (or more) part(s) of the chain, while other parts of the chain (in the intracellular and extracellular environments and close to the ionic groups of the lipid bilayer) are enriched in polar amino acids (e.g. glutamate and aspartate). In Class III proteins, for example glycophorin, a single hydrophobic domain traversing the membrane separates the two polar domains at the cytoplasmic and extracellular faces (*Figure 2.8*). In Class IV proteins, several hydrophobic sequences are separated by polar sequences that vary in length together with the two terminal polar domains. The polypeptide chain thus loops back and forth through the bilayer and, for example, can provide a polar channel similar to that depicted in *Figure 2.9*. Protein 3 of the erythrocyte membrane and various receptors and ion pumps, described in Chapters 3 and 5, are examples of Class IV proteins. *Figure 2.8* shows the arrangement of the seven transmembrane domains of bacteriorhodopsin, a light-driven proton pump present in the purple membrane of *Halobacterium halobium*.

The secondary structure of membrane proteins that span the membrane has been studied by techniques, such as infra-red and Raman spectroscopy and by circular dichroism. Most of the non-polar amino acid sequences located within the membrane have been shown to have an α-helical conformation.

If the amino acid sequence in the α-helical region of a membrane protein consists of alternating short sequences of two to three non-polar and polar amino acids, then one side of the helix displays predominantly non-polar side chains while the opposite side displays mainly polar side chains. A tetrad of such helices (see *Figure 2.9*) with their polar faces adjacent can thus provide an ion-conducting channel through the non-polar environment of the membrane.

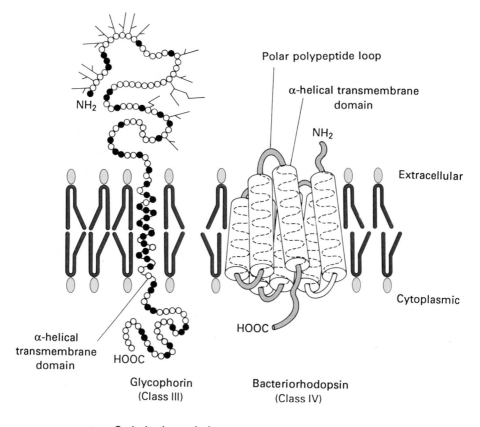

Figure 2.8. Topography in the lipid bilayer of Class III and Class IV membrane proteins.

2.2.4 Class V proteins

These proteins are attached to the membrane by a phospholipase-sensitive lipid anchor. The details of the lipid anchor have been deduced by studying glycoproteins purified from thymocytes and trypanosomes. These proteins contain at their carboxyl terminus a novel glycolipid which consists of phosphatidylinositol, ethanolamine, mannose, glucosamine and other sugars (e.g. galactose and galactosamine) that may vary (*Figure 2.10*). The glycolipid component is attached to the protein in the endoplasmic reticulum before the protein is transferred to the plasma membrane. Many ectoenzymes, that is membrane-bound enzymes, asymmetrically orientated in the plasma membrane so that their catalytic sites are exposed to the external medium are attached by a lipid anchor, for example 5′-nucleotidase, alkaline phosphatase and

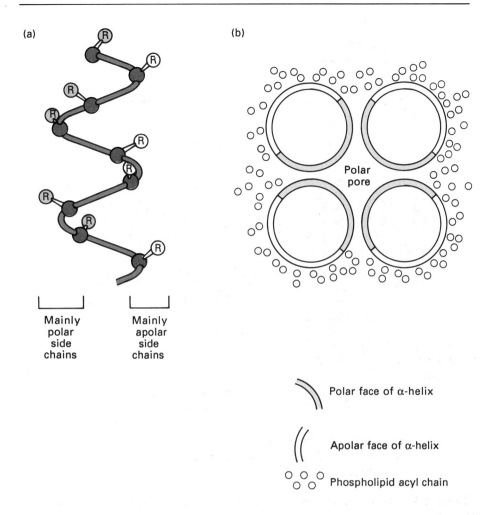

(a)

Mainly
polar
side
chains

Mainly
apolar
side
chains

(b)

Polar
pore

Polar face of α-helix

Apolar face of α-helix

Phospholipid acyl chain

Figure 2.9. Model of a polar pore. (a) Transmembrane α-helix of pore-forming protein. (b) Diagrammatic section through apolar core of membrane, parallel to surface.

acetylcholinesterase. The functional significance of the membrane glycolipid anchor is unknown. The anchor may confer a high degree of lateral mobility in the membrane, as in the case of the Thy-1 receptor found in thymocytes and brain and provide a potential for the rapid release of the receptor by an anchor-specific phospholipase.

2.2.5 Protein organization in plasma membrane domains

Proteins are able to diffuse laterally in the plane of the membrane but the rates are 100–100 000 times slower than that of lipids. Some proteins such as bacteriorhodopsin form large aggregates and are virtually immobile in the membrane. In tissues where cell surfaces are functionally polarized, there are

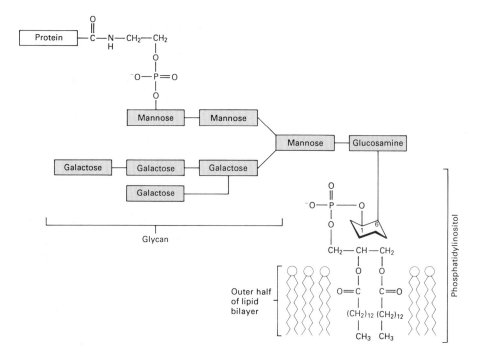

Figure 2.10.   Molecular structure of a Class V protein.

restrictions to this lateral mobility, otherwise functional polarity could not be maintained and the efficient functioning of the cell would be compromised. In most specialized cell surfaces, for example the brush border of kidney cells and intestinal cells, the plasma membrane is underlaid by an extensive cytoskeletal network. These membranes contain high amounts of 5'-nucleotidase, alkaline phosphatase and many aminopeptidases. The Na^+/K^+-ATPase ion pump on the other hand is functional in the basolateral domain of plasma membranes. Tight junctions (see Section 5) separate these surface domains. The complex rigid structure of these junctions limits the movement of protein molecules between the functional domains but apparently they permit the lateral movement of phospholipid molecules in the inner half of the lipid bilayer.

Other features which condition the lateral movement of proteins in membranes are their interactions with the cytoskeleton and the glycocalyx.

3. The glycocalyx

The predominantly carbohydrate-containing region external to the lipid bilayer is called the glycocalyx or cell coat. This layer may be as much as 50 nm

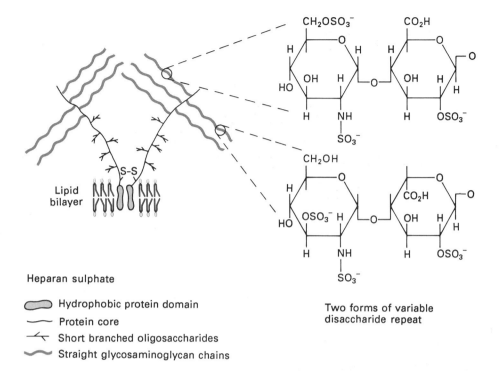

Lipid bilayer

Heparan sulphate

- Hydrophobic protein domain
- Protein core
- Short branched oligosaccharides
- Straight glycosaminoglycan chains

Two forms of variable
disaccharide repeat

Figure 2.11. Structure of heparan sulphate.

thick—much thicker than the membrane itself. In plant cells and in Gram-negative bacteria this extracellular matrix is extremely complex with the molecules being organized into a discrete outer cell wall which lends support to the internal structures. The glycocalyx is made up of the extracellular domains of integral membrane glycosylated molecules as well as other glycosylated molecules bound ionically to these domains. The molecules bound to these membrane components may be glycoproteins themselves, such as fibronectin, although the major components are proteoglycans.

Whereas the carbohydrate chains of glycoproteins are commonly no more than 7–18 residues long, proteoglycans can have as many as 40–100 residues per chain and each proteoglycan molecule may contain many carbohydrate chains. These chains, which are unbranched and commonly linked to the free hydroxyl group of serine, are referred to as glycosaminoglycan chains.

Some proteoglycan molecules (90% heparan sulphate) retain a relatively short transmembrane amino acid chain and are thus anchored into the membrane. Others interact with surface receptors through ionic interactions. *Figure 2.11* shows the membrane-bound form of heparan sulphate, illustrating the extent of the extracellular domain compared to the membrane-bound domain. The glycosaminoglycan side chains contain a repeated disaccharide unit attached to

Table 2.2. Major components of the extracellular matrix

Component	Size	Properties
Fibronectin	440 kd; monomers, 210 and 222 kd	May be membrane-associated (via integrin-receptor) or in plasma
Vitronectin	Monomer, 70 kd	Also called serum-spreading factor
Collagens	Many different types (at least 12) of varying size	Present in most connective tissues and basement membrane
Heparan sulphate proteoglycans	Protein core size range 150–400 kd. ~4 heparan sulphate chains per side chain	Interact with cell surface and other extracellular matrix components. Anionic permeability barrier in basement membrane. Implicated in presentation of growth factors to cells
Laminin	900 kd; monomers, 400 and 220 kd	Basement membrane component in epithelia and endothelia
Thrombospondin	450 kd; monomers, 150 and 185 kd	Widely distributed. Implicated in platelet aggregation

a core sugar unit. These disaccharide units often contain the oxidized forms of common sugars (uronic acids), for example glucuronic acid and amino sugars which may be *N*- or *O*-sulphated. The precise molecular structure of this repeating disaccharide may vary from chain to chain in the same molecule. *Figure 2.11*, for example, shows two forms of the disaccharide repeat of heparan sulphate.

The numerous and long glycosaminoglycan chains of these proteoglycans form the outer limits of the animal cell surface. Proteoglycan molecules are important in cell–cell adhesion, and they are involved in embryogenesis and cell proliferation. Some of the major components of the extracellular matrix and their properties are shown in *Table 2.2*.

4. The cytoskeleton

The cytoskeleton stabilizes the plasma membrane and maintains cell shape. For example, the microvillus core of the intestinal brush border contains long filaments of actin extending the length of the microvillus and they associate with the underlying terminal web. A number of proteins form bridges between the actin molecules and these include villin, fimbrin and myosin. The actin filaments of the microvillus core in turn are attached to the overlying membrane by spirally arranged bridges which are thought to consist of a 110-kd polypeptide and calmodulin.

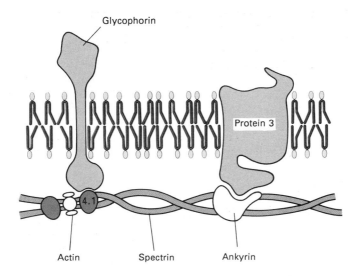

Figure 2.12. Interaction between cytoskeletal and integral proteins of the human erythrocyte membrane.

Spectrin has long been recognized as an important membrane-associated protein in the erythrocyte but non-erythroid spectrins (fodrins) have been detected in a variety of tissues and cells including intestinal brush borders, neuronal tissue, muscle cells, sperm cells and fibroblasts. Erythrocyte spectrin exists as flexible rods of about 110 nm in length and are made up of 240-kd and 225-kd subunits arranged as a repeating triad of helical segments which can associate to form large oligomeric structures.

Two principal proteins [ankyrin and protein 4.1 (*Figure 2.12*)] link the cytoskeletal framework to the overlying red cell membrane, with ankyrin forming a bridge between spectrin and the integral membrane protein, protein 3. Protein 4.1 links spectrin to a glycophorin, and it is thought to stabilize the actin – spectrin association. In some forms of haemolytic anaemia (elliptocytosis and spherocytosis) the lack of protein 4.1 or spectrin results in abnormal erythrocyte shape.

5. Intercellular junctions

The surface of cells organized into tissues and organs is not uniform, but is instead punctuated by a variety of intercellular junctional complexes (*Figure 2.13*). Desmosomes are junctions that attach cells tightly to one another; they are also the membrane areas where the cytoplasmic tonofilaments are attached to the plasma membrane, helping to determine cell shape and polarity. Desmosomes are especially prevalent in tissues subject to mechanical stress, for example skin, and in organs carrying out mechanical work, for example heart. They are complex

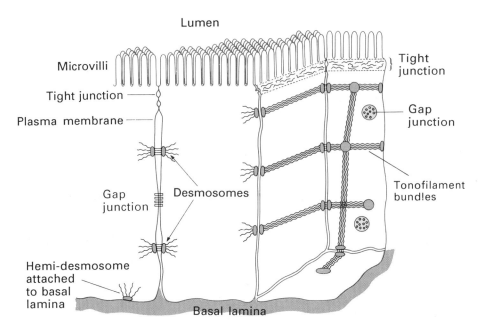

Figure 2.13. Arrangement of various categories of intercellular junctions in columnar epithelial cells.

structures constructed of several proteins, for example desmoplakins and various glycoproteins attaching the cells strongly together.

Tight junctions are found mainly in absorptive and secretory epithelial cells where they demarcate the plasma membrane into apical and basolateral domains. They control the movement of small ions and molecules through the extracellular space between cell layers. In brain capillaries, for example, they control the direct entry of metabolites, drugs, etc. from the blood into the brain. Freeze-fracture electron microscopy reveals a pattern of intramembranous particles arranged as interdigitating strands and grooves. At the molecular level, little is known currently of their component proteins and their arrangement in the lipid bilayer.

Two categories of intercellular junctions feature in facilitating cell–cell communication. Gap junctions (see Section 2 of Chapter 5) are widely distributed in animal tissues where they ensure tissue homeostasis, development processes, etc. Synaptic junctions ensure directional communication in neuronal cells and tissues. The pre-synaptic membrane of the neuron is involved in the release of neurotransmitters from synaptic vesicles into the synaptic cleft. Receptors located on the post-synaptic membrane, for example muscarinic and γ-aminobutyric acid (GABA), respond in an excitatory or inhibitory manner to the neurotransmitters ensuring the degree of onward transmission of the electrical response. Many proteins are located in the post-synaptic junctions, including proteases, phosphorylases and Ca^{2+} and calmodulin-dependent protein kinases.

6. Further reading

General (especially lipids)

Jain,M.K. (1988) *Introduction to Biological Membranes.* 2nd edition. John Wiley & Sons, New York.
Datta,D.B. (1987) *A Comprehensive Introduction to Membrane Biochemistry.* Floral Publishing, Madison.

Cholesterol

Yeagle,P.L. (1985) *Biochim. Biophys. Acta,* **822**, 267–288.

Phospholipids

Hawthorne, J.N. and Ansell,G.B. (1982) *Phospholipids.* Elsevier, Amsterdam.

Ether lipids

Mangold,H.K. and Paltauf,F. (1983) *Ether Lipids—Biochemical and Biomedical Aspects.* Academic Press, New York.

Glycolipids

Curatolo,W. (1987) *Biochim. Biophys. Acta,* **906**, 111–136.
Curatolo,W. (1987) *Biochim. Biophys. Acta,* **906**, 137–160.

Dolichol

Chojnack,T. and Dallner,G. (1988) *Biochem. J.,* **251**, 1–9.

Protein structure

Eisenberg,D. (1984) *Annu. Rev. Biochem.,* **53**, 595–623.

Membrane domains

Evans,W.H. (1980) *Biochim. Biophys. Acta,* **604**, 27–64.
Simons,K. and Fuller,S.D. (1985) *Annu. Rev. Cell Biol.,* **1**, 243–288.

Proteoglycans

Gallagher,J.T., Lyon,M. and Steward,W.P. (1986) *Biochem. J.,* **236**, 313–325.
Poole,A.R. (1986) *Biochem. J.,* **236**, 1–14.

Cytoskeleton

Mooseker,M.S. (1985) *Annu. Rev. Cell Biol.,* **1**, 209–241.
Bennett,V. (1984) *Annu. Rev. Biochem.,* **54**, 273–304.

Junctions (general)

Junctional Complexes of Epithelial Cells. (1987) *Ciba Foundation Symposium 125.* John Wiley and Sons, Chichester.

Desmosomes

Garrod,D.R. (1986) *J. Cell Sci. Suppl.,* **4**, 221–237.

Tight junctions

Gumbiner,B. (1987) *Am. J. Physiol.,* **253**, C749–C758.

Lipid anchors

Ferguson,M.A.J. and Williams,A.F. (1988) *Annu. Rev. Biochem.,* **57**, 285–320.
Sefton,B.M. and Buss,J.E. (1987) *J. Cell Biol.,* **104**, 1449–1453.

Low,M.G. and Saltiel,A.R. (1988) *Science,* **239**, 268–275.

3

Membrane receptors

---

1. General properties of membrane receptors

Receptors constitute one of the most important categories of membrane proteins. In most instances, they are glycoproteins functioning at the cell surface, but receptor 'pools' are also held in reserve in intracellular membranes. Receptors are crucial for transducing a wide range of external stimuli and signals (e.g. light, smell, hormones, drugs, growth factors) into a co-ordinated physiological response. Transduction of the signal often involves the activation of key membrane-bound enzymes and the generation of various 'second messengers'.

The functions attributed to membrane receptors continues to increase. The receptor concept, first introduced by pharmacologists to account for the specific actions on cells of neurotransmitters and drugs, and more recently extended to explain the mechanism of action at the cell surface of polypeptide and other hormones, has been widened further. Receptors control the permeability of the plasma membrane by opening and closing ion channels. Others function as transport proteins, as illustrated by the delivery by a low-density lipoprotein receptor of its ligand to lysosomes, the uptake of iron, bound to transferrin by the transferrin receptor and the transcellular itinerary from blood to body spaces, for example intestinal lumen, of the polymeric immunoglobulin A bound to its receptor – secretory component. The integrin family of receptors link the extracellular matrix components with the cytoskeleton and these and other receptors, for example cadherins at the cell surface, exercise crucial control on many aspects of cell – cell interactions. Even pathogens gain entry into cells by exploiting their ability to identify and combine with receptors at the cell surface. Thus a wide range of enveloped viruses, bacterial toxins, parasites and plant-derived lectins such as ricin and abrin take advantage of endocytic pathways to gain entry into cells. It is not surprising therefore that receptors have been classified in many ways, for example according to their structure, the nature of the ligand bound, the mechanism of action and the metabolic consequences (*Table 3.1* and *Figure 3.1*).

Most receptor proteins traverse the membrane once or many times. Exceptions to this general rule are being discovered. For example the Thy-1 receptor on thymocytes and brain cells is anchored into the membrane by a phosphoinositide tail (see Section 2.2.4 of Chapter 2). Three major domains—external,

31

Table 3.1. Classification of cell surface receptors

Receptor type	Nature of ligand
Receptors linked to ion channels Na$^+$ Cl$^-$ Cl	Nicotine acetylcholine γ-Aminobutyric acid (GABA) Glycine
Receptors with protein kinase activity	Epidermal growth factor (EGF) Insulin Platelet-derived growth factor (PDGF)
Receptors that *activate* adenylate cyclase	β-Adrenergic Vasopressin Glucagon Thyroid-stimulating hormone (TSH) Histamine Adenocorticotrophic hormone (ACTH) Prostacyclin Parathyroid hormone
Receptors that *inhibit* adenylate cyclase	α-Adrenergic Muscarinic acetylcholine Prostaglandin E$_1$ Adenosine Opiate
Receptors that activate phosphoinositide hydrolysis (phospholipase C)	α-Adrenergic Muscarinic acetylcholine Substance P Gonadotrophin-releasing hormone Angiotensin Thyrotrophin-releasing hormone Thromboxane A$_2$
Receptors that activate phospholipase A	Histamine Vasoactive intestinal peptide Bradykinin
Receptors that modulate the immune response	Various antigens, antibodies and lymphokines
Receptors linked to the extracellular matrix (Integrins)	Fibronectin
Receptors that function as delivery systems	Transferrin Asialoglycoproteins Low-density lipoproteins

transmembrane and cytoplasmic—are demarcated as shown in *Table 3.2*. The topography of the amino- and carboxyl-termini also varies, this being a direct consequence of the mechanism of synthesis (see Section 2.1 of Chapter 4).

Table 3.2. General structure of cell surface receptor proteins

Domain	Properties
Extracellular	Contains ligand binding site
	Usually comprises most of protein
	Contains complex and high mannose oligosaccharides
	May contain cysteine-rich sequences
	Location of sequence homology between receptor families
Transmembrane	Short ($\sim$20 amino acids) containing mainly hydrophobic residues
	May be arranged as one or many membrane-spanning α helices
	Highly conserved in receptors linked to G-proteins
Cytoplasmic	Involved in signal transduction and receptor trafficking
	Length varies
	Region where phosphorylation occurs
	Intrinsic protein kinase activity may reside here

2. Receptors and signal transduction

The plasma membrane of animal cells contains biochemical mechanisms that translate the complex repertoire of extracellular stimuli into chemical, ionic and electrical intracellular events.

2.1 Receptors with intrinsic kinase activity

2.1.1 Insulin receptor

Most animal cells possess insulin receptors, with the number on the cell surface ranging from 100 to 100 000. Within seconds of the binding of insulin to a receptor on the plasma membrane, the receptor autophosphorylates. Minutes later, the receptor – insulin complex is internalized and glucose, amino acid and ion transport are amongst the many ensuing metabolic changes; these processes are a prelude to the long term effects of insulin on cell metabolism and the growth of target cells.

The structure in the plasma membrane of the insulin receptor is shown in *Figure 3.2*. It is a transmembrane glycoprotein constructed of two α subunits (130 kd) and two β subunits (95 kd) connected by disulphide bridges. An immediate consequence of the binding of insulin to the receptor is the autophosphorylation of tyrosine residues at the cytoplasmic aspect of the β subunit by the receptor's intrinsic tyrosine phosphokinase. Indeed autophosphorylation is functionally a feature of certain classes of receptors and has been shown with a number of growth factor receptors, including epidermal growth factor (EGF), platelet-derived growth factor (PDGF), and insulin-like growth factor (IGF-1) receptors.

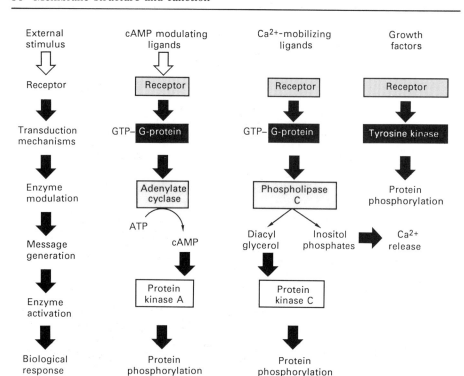

Figure 3.1.   Summary of events cascading at the plasma membrane and inside the cell after activation of various receptor categories by external stimuli.

Analysis of the roles played by the various domains of the insulin receptor is being pursued. Insulin-binding, disulphide cross-linked, transmembrane, tyrosine kinase and cytoplasmic tail domains of the protein can be categorized. To assign to each of these domains the various aspects of receptor functioning and to assess their importance, the insulin receptor of cultured cells has been mutated yielding various modified and truncated forms. For example, removal of the domain containing the intrinsic tyrosine kinase activity involved in receptor autophosphorylation results in a loss of biological activity and endocytosis of the receptor is blocked. It is concluded that the effects of insulin binding are probably mediated by the tyrosine kinase activity of the receptor.

Unlike many other hormone receptors, the insulin receptor appears in our present understanding to function without the involvement of known second messengers. The molecular events linking activation of the insulin receptor to changes in phosphorylation of target enzymes are still far from being understood, but it was recently shown that the binding of insulin to its receptor promotes the hydrolysis of a phosphatidylinositol glycan containing glucosamine and other sugars. The two products, inositol phosphoglycan and diacylglycerol, may selectively regulate protein kinases and phosphodiesterases discussed below.

The mechanism by which insulin stimulates glucose transport in many cells

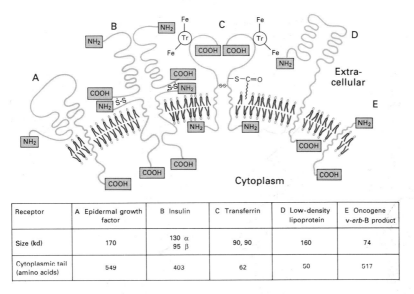

Receptor	A Epidermal growth factor	B Insulin	C Transferrin	D Low-density lipoprotein	E Oncogene v-erb-B product
Size (kd)	170	130 α 95 β	90, 90	160	74
Cytoplasmic tail (amino acids)	549	403	62	50	517

Figure 3.2.   Proposed topographical arrangement of receptors and an oncogene product in the plasma membrane. The v-*erb*-B oncogene product is a homologous truncated form of the EGF receptor. All receptors are glycosylated in their extracellular domain. The insulin and transferrin receptors are homodimers. The transferrin receptor contains an acylated fatty acid moiety. Note the different orientation of the transferrin receptor.

involves its acceleration of the movement to the plasma membrane of glucose transporters held in an intracellular pool of small vesicular structures. The glucose transporter of adipocytes is a 42-kd membrane protein, and in models of sugar transporters in animal cells and bacteria the molecule traverses the lipid bilayer 12 times, with both amino- and carboxyl-termini located at the cytoplasmic side.

2.1.2 Growth factor receptors

Receptors for growth factors also incorporate into their structure an intrinsic protein (tyrosine) kinase. Twenty or more well defined growth factors are known which stimulate cell proliferation and differentiation. The best studied is the receptor for epidermal growth factor (EGF)—a mitogen constructed of 53 amino acids.

The EGF receptor structure, determined in a carcinoma cell line, that over-expresses the receptor 20- or 50-fold, is shown in *Figure 3.2*. It consists of a single polypeptide of 1186 amino acids. The extracellular domain of 621 amino acids has two notable structural features—a large amount of cysteine resulting in the presence in the intact receptor of up to 25 disulphide bridges making the receptor highly resistant to proteolysis and a large amount of N-linked carbo-hydrate. The cytoplasmic domain contains the tyrosine phosphokinase activity. An interesting feature of the receptor structure also found in other growth factor receptors is the presence of a stretch of 13 residues enriched in basic amino acids.

Table 3.3. Properties of some cell surface receptors

Ligand	Cell type studied	Receptor mass (kd)	Location of NH$_2$-terminus	Receptor Recycling time (min)	Receptor half-life (h)
Low-density lipoproteins	Fibroblasts	160	Extracellular	6	25
Asialoglycoproteins	Hepatocytes	42–54	Cytoplasmic	8	20
Transferrin	Cultured hepatocytes	90	Cytoplasmic	10	>8
Insulin	Adipocytes	130 (α) 95 (β)	Extracellular	10	10
Epidermal growth factor	Fibroblasts	130	Extracellular	–	1

This highly basic region is thought to function as a 'stop transfer' sequence after the insertion of the amino-terminus end across the endoplasmic reticulum in its biosynthesis. Charged amino acid sequences are emerging as a general feature of cytoplasmic loops in transmembrane proteins. As shown in *Table 3.3* the half-life of the EGF receptor is much shorter than for many other receptors indicating that it does not undergo recycling between the plasma membrane and the endocytic compartment in A-431 cells but is transferred to lysosomes and degraded.

2.1.3 Protein kinases

These are a class of enzymes found in all eukaryotes that feature in intracellular signalling by reversible phosphorylation of substrates, many of which are membrane-bound. At least three categories of kinases have been identified, namely cyclic nucleotide-dependent, Ca^{2+}/calmodulin-activated and protein kinase C. They all have a striking similarity in amino acid sequence at their catalytic sites. However, minor differences exist between the structure of classes of protein kinases that phosphorylate at serine and threonine and those that phosphorylate at specific tyrosine residues.

Many protein–tyrosine kinases are themselves receptors that function as signal transducers for circulating peptide hormones and growth factors. For example, the insulin and growth factor receptors exhibit tyrosine kinase activity that is central to their role in the transmission of information into the cell. A unique feature of serine-/threonine-specific protein kinase C is its dependence on phospholipids and Ca^{2+} and its activation by a second messenger, diacylglycerol (Section 2.3). Indeed, the protein kinases through their promiscuous phosphorylation are centrally involved in regulating many aspects of cell behaviour, a function aided by their dual membrane and cytosolic location in cells.

Agents such as phorbol esters that stimulate protein and steroid secretion and are tumour-promoting react specifically with protein kinase C. Unravelling the precise roles of protein kinases in cell signalling is difficult because a large number of enzymes and substrates in and associated with membranes have been identified. The kinases themselves are often multisubunit enzymes with catalytic and regulatory subunits. Phosphorylation of specific amino acid residues, for example tyrosine, may induce a conformational change in the topography of the membrane constituent that can influence positively or negatively the level of signal transmission across the membrane into the cell. Protein kinases act in concert with phosphatases in the membrane that dephosphorylate proteins thus allowing regulatory circuits to be established. The target proteins and enzymes phosphorylated by protein kinases are being extensively investigated.

2.1.4 Oncogenes and membrane receptors

Modifications to growth factors and their receptors are linked to oncogenesis and the subversion of cell behaviour. Of the oncogene products identified thus far, some are related to membrane receptors in two respects. First, many products of the oncogenc family (c.g. *src*, v-*crb*) arc tyrosinc kinascs locatcd at the plasma membrane. Second, strong evidence indicates that certain retroviral-transforming genes have originated from genes encoding growth factor receptors. For example, comparison of the complete amino acid sequence of human epidermal growth factor receptor and that of the sarcoma-inducing avian erythroblastosis retrovirus transforming gene v-*erb* has revealed similarit*j*. Indeed, the receptor-related oncogenes are classed as a family because they possess tyrosine kinase activity and are associated mainly with the cytoplasmic aspect of the plasma membrane (*Figure 3.2*).

2.2 Receptors generating cyclic nucleotides

This mechanism of signal transduction was one of the earliest discovered. Glucagon and adrenalin, after binding to their respective receptors, stimulate adenylate cyclase to produce cyclic AMP (cAMP), that, in turn, phosphorylates key enzymes involved in glycogen metabolism (e.g. glycogen synthase, phosphorylase) and lipolysis. Other receptors (see *Table 3.1* for a comprehensive list) inhibit cAMP production allowing a fine balancing of the signalling systems. The effects of hormones can vary between tissues. For example, glycogen metabolism in skeletal muscle responds mainly to adrenalin whereas that of liver responds mainly to glucagon. Connecting these receptors to the adenylate cyclase are the G-proteins described below.

2.3 Receptors that influence phosphoinositide turnover

Many receptors influence cell behaviour by hydrolysis of a minor phospholipid, phosphatidylinositol-4,5-bisphosphate. Phosphoinositidase-specific phospholipase C hydrolyses this phospholipid to produce inositol-1,4,5-trisphosphate and diacylglycerol, both of which act as second messengers (*Figure 3.1*).

Inositol-1,4,5-trisphosphate (IP_3) is water-soluble and after its release into the cytoplasm, it binds to an intracellular membrane-bound Ca^{2+} store. The localization of this Ca^{2+} store is uncertain. It is either a part of the endoplasmic

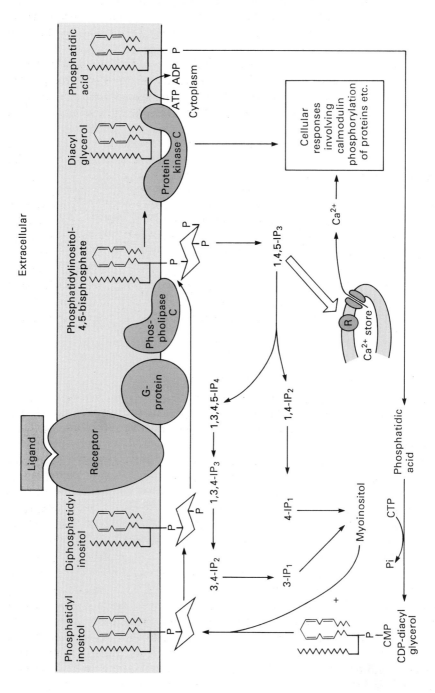

Figure 3.3.   Signal transduction and the phosphoinositide cycle. R is the IP$_3$ receptor on membranes encasing intracellular Ca^{2+} stores.

reticulum or an uncharacterized membrane. IP_3 binding promotes Ca^{2+} efflux through a channel thereby modifying the activities of many enzymes, especially those activating glycogen breakdown in liver and contraction in muscle tissue. The further metabolism of IP_3 is shown in *Figure 3.3*. Ultimately, all phospho-inositides are hydrolysed to inositol.

The other product of phosphatidylinositol-4,5-bisphosphate breakdown is diacylglycerol. It is retained in the plasma membrane where it activates protein kinase C (Section 2.1). Phorbol esters, a group of natural products produced from croton oil and potent promoters of skin tumours, bind to protein kinase C in a similar fashion to diacylglycerol.

Diacylglycerol is hydrolysed further by lipase to yield phosphatidic acid and arachidonic acid, with the latter product being metabolized to yield the eicosanoids. This is a generic name for compounds derived from C-20 poly-unsaturated fatty acids, but mainly arachidonic acid. The eicosanoid group in-cludes the prostaglandins, thromboxane and leukotrienes, all compounds with potent biological activities produced mainly by white blood cells and are involv-ed in inflammatory and hypersensitivity reactions. In the presence of ATP, diacylglycerol yields phosphatidic acid that is converted to phosphatidylinositol (*Figure 3.3*). This is further phosphorylated in the plasma membrane to yield phosphatidyl-4,5-bisphosphate, thereby completing the phosphatidylinositol cycle. GTP-binding (G) proteins are implicated in the transmission of a signal produced by the formation of a ligand – receptor complex, and their precise nature will soon be established.

2.4 Receptors incorporating or modulating ion channels

The plasma membrane maintains and controls gradients of ions between the cell interior and the outside. This involves pumps (discussed in Chapter 5) and channels that are either receptors in their own right or are indirectly controlled by receptors.

The nicotinic acetylcholine receptor is an example of a multisubunit receptor (*Figure 3.4*). Binding of acetylcholine to the receptor – agonist binding site directly opens the channel for a few milliseconds. The inward movement of Na^+ at the neuromuscular junction results in an action potential. In contrast, binding of ligand to the muscarinic acetylcholine receptor affects the conductance of ion channels, with a G-protein implicated as an intermediary (*Figure 3.8*).

A further example of a receptor directly controlling ion movements is provided by the GABA (γ-aminobutyric acid) receptor shown in *Figure 3.5*. The role of GABA in synaptic transmission in the mammalian central nervous system is more firmly established than that of any other neurotransmitter. The receptor structure provides a pore through which chloride ions pass into the cell thereby raising the membrane potential. The cell thus becomes more refractory to stimulation, hence explaining the role of GABA as an inhibitory neurotransmitter. Tran-quillizers (e.g. benzodiazepines) bind to this receptor and potentiate the effects of GABA and thus increase the permeability of the chloride channel.

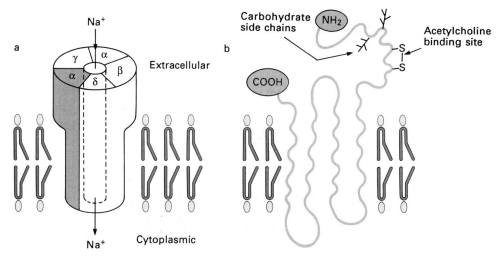

Figure 3.4. Model of the nicotinic acetylcholine receptor. (**a**) shows the receptor to be constructed of two α, and single β, γ and δ subunits each contributing to the channel wall. A proposed topographical arrangement of the α subunit is shown in (**b**).

2.5 Receptors coupled to G-proteins—general structural features

The structure of the β-adrenergic and muscarinic receptors are shown in *Figure 3.6* and also included for comparison is an alternative depiction to that shown in *Figure 3.10* of the bovine rhodopsin receptor. These receptors are coupled to G-proteins and the structural features in each case are the seven membrane-spanning sequences.

Knowledge of the structure of families of receptors exemplified by those coupled to G-proteins shows that they are products that have evolved from an ancestral gene by duplication. During evolution, individual receptors have retained general structural features such as the mechanism of anchoring the protein in the lipid bilayer while adapting to allow response to external stimuli that vary from light to neurotransmitters, and to communicate specific signals via the mediation of G-proteins to the effector systems located within and on the cytoplasmic aspect of the plasma membrane.

2.5.1 G-proteins—a family of signal transducers

The G-proteins are a growing family of guanyl nucleotide-binding (G) regulatory proteins. They are found associated mainly with the plasma membrane although recently they have also been found in Golgi and endocytic membranes. As shown in *Table 3.4* G-proteins are implicated in a diverse range of physiological processes.

G-proteins are built to a common design. They are heterotrimeric molecules consisting of a major α subunit of variable molecular weight (39–52 kd) to which

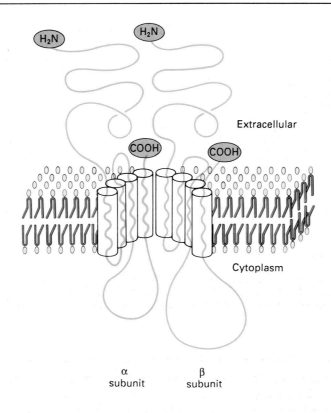

Figure 3.5. Model of the GABA subunit based on its construction from α and β subunits, although other subunits may also contribute to its structure.

Table 3.4. Examples of signal transduction by receptors interacting with G-proteins

Receptor	Stimulus	Effector	G-protein	Examples of responses
β-Adrenergic	Adrenalin	Adenylate cyclase	G_s	Glycogen breakdown Acceleration of heart rate
Serotonin	Serotonin	Adenylate cyclase	G_s	Widespread, e.g. neuronal excitability
Glucagon	Glucagon	Adenylate cyclase	G_s	Glycogen breakdown
Rhodopsin	Light	cGMP phospho-diesterase	Transducin	Visual excitation
Mast cell IgE	IgE-antigen	Phospholipase C	?	Secretion
Muscarinic subtypes	Acetyl-choline	K$^+$-channel Phospholipase C Adenylate cyclase	? ? G_i	Slowing of heart rate Excitation or inhibition of central neurons

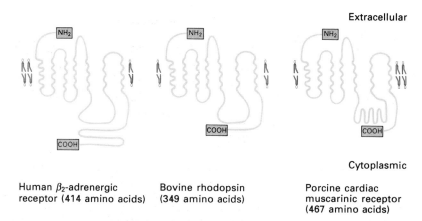

Figure 3.6. Structure of three receptors constructed of seven membrane-spanning domains that signal via the intermediation of G-proteins. The overall structures are similar although the extent of amino acid identity is low.

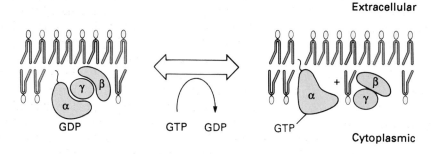

Figure 3.7. The GTP–GDP cycle of G-proteins. The α-subunit of G-proteins is anchored to the membrane by a myristic acid tail.

GTP binds and two smaller and tightly coupled subunits [(the β (36 kd) and γ (8 kd) chains]. As shown in *Figure 3.7*, activation of the receptor results in GDP bound to the α subunit being replaced by GTP. Subsequently, GTP hydrolysis by a guanine triphosphatase, which is an integral component of the α subunit, deactivates the system. The β chain subunits of G-proteins obtained from a variety of sources are nearly identical, whereas the α subunit contains a myristic acid 'tail' that helps to anchor it in the membrane.

The most studied G-proteins are those that stimulate (designated G_s) or inhibit (G_i) the adenylate cyclase and thus the production of cAMP, the response being dictated by the nature of the receptor involved.

Of great help in the study of G-proteins has been the observation that toxins (e.g. cholera and pertussis) interfere with their role in signal transduction. Most plant and bacterial toxins have a similar general structure, consisting of two polypeptide chains linked by disulphide bridges. It is now possible to explain

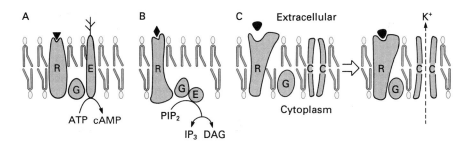

Figure 3.8. Mechanisms of action of receptors linked to G-proteins. (**A**) Stimulation of adenylate cyclase (E) by a receptor (R) (e.g. glucagon, β-adrenergic) involving G_s. (**B**) Stimulation of phospholipase C (E) by α-adrenergic receptor to yield IP_3 and diacylglycerol (DAG). (**C**) Stimulation of the muscarinic receptor (R) in cardiac cells by acetylcholine leads to a G-protein-mediated rearrangement resulting in a K^+ channel (C) opening.

at the molecular level the pathogenesis of toxin-induced diseases. The mechanism of action of cholera toxin is shown in *Figure 3.9*. The increased cAMP production induced by the toxic binding to ganglioside on the plasma membrane of intestinal cells causes enhanced secretion of salt and H_2O into the gut; this dehydration can prove fatal. Other enterotoxins secreted by bacteria act by a similar mechanism. For example, pertussis toxin which ADP-ribosylates G_i causes diverse pathological effects including sensitization, the histamine response, enhanced insulin secretion, and it acts as a T cell mitogen.

Other members of the G-protein family which function in signal transduction are being discovered. Three variants of G_i are known; G_o is implicated in the control of K^+ channels and is abundant in the brain. Transducin functions in vision (see Section 2.6). G-proteins are involved in phosphoinositide turnover leading to activation of protein kinase C and calcium mobilization (Section 2.3) and in the coupling of neurotransmitter receptors to ion channels without direct involvement of second messengers (*Figure 3.8*). Indeed, the implication of G-proteins in phosphoinositide turnover and intracellular Ca^{2+} mobilization is one of the most significant unifying themes in receptor biochemistry.

The *ras* oncogene product p.21 (a 21-kd polypeptide) shares many properties with G-proteins, for it also binds GTP, shows GTPase activity and is associated with the inner surface of the plasma membrane. The products of the *ras* oncogene also show some amino acid sequence homologies with transducin. Mutation of *ras* or overexpression of p.21 lead to a malignant phenotype. GTP-binding proteins are also involved in secretion. Using temperature-dependent yeast mutants defective in constitutive secretion, it has been shown that a 23.5-kd *ras*-like protein is necessary. These GTP-binding proteins are localized in the Golgi apparatus.

2.6 Receptor mechanisms in vision and olfaction

In the eye, pigmented sensory cells transform a light stimulus into an electrical signal. As shown in *Figure 3.10*, rod cells contain a stack of photosensitive discs,

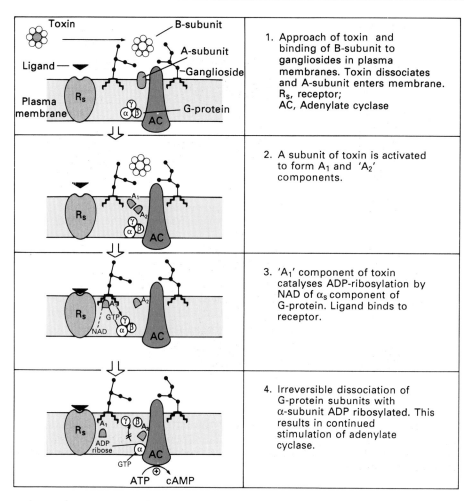

	1. Approach of toxin and binding of B-subunit to gangliosides in plasma membranes. Toxin dissociates and A-subunit enters membrane. R_S, receptor; AC, Adenylate cyclase
	2. A subunit of toxin is activated to form A_1 and 'A_2' components.
	3. 'A_1' component of toxin catalyses ADP-ribosylation by NAD of α_S component of G-protein. Ligand binds to receptor.
	4. Irreversible dissociation of G-protein subunits with α-subunit ADP ribosylated. This results in continued stimulation of adenylate cyclase.

Figure 3.9.   Molecular mechanism of action of cholera toxin in the stimulation of adenylate cyclase activity at the intestinal cell plasma membrane.

surrounded by a plasma membrane and embedded in the disc membrane are the rhodopsin molecules. Absorption of a single photon by rhodopsin can close about 1000–3000 sodium channels per second.

Rhodopsin, the photoreceptor protein in the discs, traverses the membrane seven times. It contains a chromophore-retinal attached by a Schiff base to a lysine residue attached in the centre of the bilayer to the seventh transmembrane helix. On absorption of photons, the retinal isomerizes from the *cis* to the *trans* form via a series of intermediates, one of which interacts with a G-protein (transducin) catalysing the exchange of bound GDP for GTP (*Figure 3.8*). The GTP-containing G-protein then activates a phosphodiesterase that hydrolyses cGMP to GMP. High cGMP levels in the cell maintain the plasma membrane

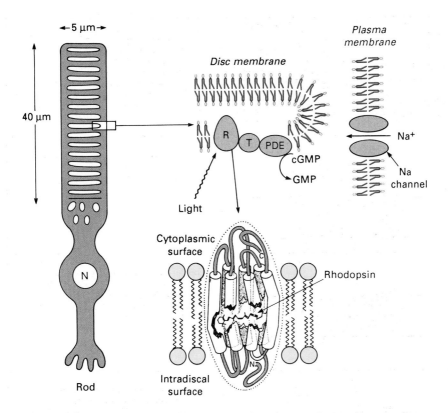

Figure 3.10. Signal transduction in vertebrate rods. The disc membranes convert light into an electrical message involving a receptor (R) incorporating rhodopsin, transducin (T) and a phosphodiesterase (PDE) hydrolysing cGMP. A model of the arrangement of the rhodopsin receptor is shown.

Na$^+$ channel in an open configuration, and light produces hyperpolarization resulting in cGMP hydrolysis and closure of the Na$^+$ channels.

The light transducing opsins of bacteria and mammals have been investigated extensively. The most thoroughly documented example is bacteriorhodopsin (described in Section 2.2.3 of Chapter 2). All opsins have similar shape and arrangement in the membrane, although there is little direct amino acid sequence homology between animal and bacterial opsins. The opsins belong to a family of receptors with a similar overall structure or 'motif' and are coupled to the guanine nucleotide regulatory (G) proteins (*Figure 3.6*), although it is to be noted that the bacteriorhodopsin which acts as a proton pump is not coupled to G-proteins.

The olfactory system discriminates between sub-nanomolar concentrations of foreign molecules in the environment. Olfactory receptor cells located in the nasal epithelium are bipolar neurons with a dendritic projection containing a group of 5 – 20 chemo-sensory cilia in contact with the mucus in the olfactory mucosa.

Olfactory cilia contain in their plasma membrane many of the essential components required for signal transduction. In view of the wide range of odorants, many categories of olfactory receptor proteins, and possibly more than one mechanism of translating the chemical information into an electrical signal, exist. Present evidence suggests that odorant receptors activate a stimulatory G-protein (G_s) thereby increasing cAMP synthesis which then activates sensory ion channels leading to depolarization of the neuronal membrane.

3. Ligand-delivering receptors

Many cell surface receptors function as membrane-embedded carriers of their bound ligands. In this section, the structure and properties of receptors involved in iron and lipoprotein transport into the cell are described. The structure of the receptor for polymeric IgA is described in Section 4, and its itinerary in the cell in Section 5.3. of Chapter 4.

3.1 The transferrin receptor

Transferrin is a 80-kd serum glycoprotein which transports iron to cells. Most animal cells possess on their cell surface high affinity receptors to which transferrin binds and the number of these receptors can vary according to the growth rate of the cell. The transferrin receptor is a dimeric glycoprotein composed of two identical subunits (*Figure 3.2*). Human hepatoma cells possess about 50 000 transferrin binding sites on their surface, and a further 100 000 in intracellular locations. After binding their ligands, transferrin receptors cluster in coated pits, are internalized into the endocytic compartment and then return to the cell surface; this itinerary takes about 10 min to complete. While ligand complexes are in general dissociated in the low pH of the endocytic compartment (see Section 5.2 of Chapter 4), the transferrin receptor – ligand complex remains intact, but the bound iron is released. After returning to the cell surface, transferrin dissociates from the receptor, further transferrin with bound iron attaches to the receptors, thereby facilitating a means for further rounds of iron transport into the endocytic compartment and leading to its translocation into the cytosol where the iron is stored (e.g. as ferritin or haemosiderin).

3.2 The low-density lipoprotein (LDL) receptor

Cholesterol is transported mainly as cholesterol ester in the serum from the liver to peripheral tissues in low-density lipoprotein (LDL). The LDL is taken up by the cells of peripheral tissues by endocytosis. The LDL receptor on the surface of these cells recognizes specific apoprotein(s) on the surface of the LDL.

The LDL-receptor is a 160-kd integral membrane glycoprotein constructed of five domains (excluding the spliced amino-terminal signal sequence). These (*Figure 3.2*), described sequentially in an extracellular to cytoplasmic direction, are:

(i) the LDL-binding sequence of about 300 amino acids (homologous to complement components C8 and C9);

(ii)   a 400-amino acid sequence homologous to the precursor for EGF which
        plays a role in recycling;
(iii) a 58-amino acid sequence rich in serine and threonine residues bearing
        O-linked sugar chains which may target the receptor to the correct plasma
        membrane domain;
(iv)   a 22 – 25-amino acid transmembrane sequence; and
(v)   a 50-amino acid cytoplasmic sequence which probably targets the receptor
        to coated pits on the plasma membrane, and onwards into the endocytic
        compartment.

After the receptor and LDL dissociate, the receptor returns to the surface,
whereas the LDL molecules are transferred to lysosomes where cholesterol esters
are hydrolysed. Cholesterol levels in the cell control not only its *de novo* synthesis
but also expression of the LDL receptor at the plasma membrane.

In familial hypercholesterolaemia the raised level of serum cholesterol is caused
by a defect in LDL uptake by receptors, exacerbated by raised *de novo* synthesis.

The frequent occurrence of genetic mutations which disturb the structure and
function of the LDL receptor is a particular feature of this disease. There are
four types of mutation which have been recognized and which have been
instrumental in revealing the molecular mechanisms regulating:

(i)   synthesis of the receptor;
(ii)   its transport to the cell surface;
(iii) its ability to bind LDL; and
(iv)   the association of the receptor with clathrin in coated pits.

These modifications have also proved useful in defining the functional importance
of the various receptor domains.

4. Receptors of the immune system and the immunoglobulin superfamily

A number of other cell surface components involved in intercellular recognition
possess a similar immunoglobulin-like overall structure. Inclusion in the immunoglobulin 'superfamily' requires the presence on the extracellular side of the
receptor of characteristic C and V amino acid domain folds with a similar three-
dimensional structure with two β sheets forming a sandwich-like arrangement
stabilized by a conserved disulphide bond. In the β strands, hydrophobic and
hydrophilic amino acids alternate, with the hydrophobic side chains orientated
to the interior of the sandwich.

The structures of a secretory component [the receptor for polymeric immunoglobulin A (IgA)], N-CAM, Thy-1 and T cell antigen receptors are shown in
Figure 3.11. Two of these receptors contain a short cytoplasmic tail, but the
receptor for IgA contains an extended cytoplasmic arm and this may be related
to its role in transporting its ligand from one pole of an epithelial cell to another,

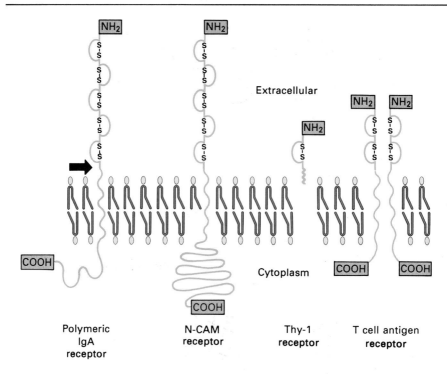

Figure 3.11.   Examples of the immunoglobulin superfamily of receptors. Polymeric IgA function is discussed in Chapter 4, Section 5.3; arrow points to position of enzymatic cleavage to produce a secretory component released into body fluids. N-CAM belongs to the cadherin family, and Thy-1 receptor is attached to the membrane by a lipid anchor. The β-sheet folding patterns held together by disulphide-bonds which characterize this family of receptors are indicated.

involving transport across several membrane boundaries. A further member of the immunoglobulin superfamily, the Thy-1 receptor, lacks even a trans-membrane amino acid sequence for it is anchored into the membrane by a phosphoinositide tail.

The members of this family continue to increase and now include the neural cell adhesion molecule receptor N-CAM (see Section 6), two myelin-associated glycoproteins, the PDGF receptor, colony-stimulating factor (CSF) I receptor, the mouse macrophage F_c receptor and a number of T cell antigen receptors, including the receptor for the human immunodeficiency virus (CD-4).

The immunoglobulin superfamily receptors can also be classified according to their function. The T cell, PDGF, CSF and Thy-1 receptors play a role in mitogenesis and cell activation. The major histocompatibility complex (MHC) receptors regulate cell interactions, and the N-CAM receptors cell adhesion. The immunoglobulin superfamily provide another example of the elaboration through evolution of a consensus structure in which diversity of amino acid sequence allows specific templates for recognition of various ligands but the structure also provides for receptor stability, thereby limiting proteolysis.

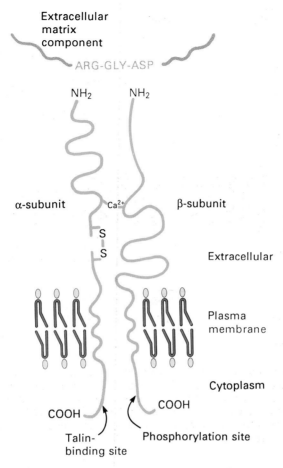

Figure 3.12.   General structure of integrin receptors. Electron microscopy has shown them to be medusa-like bodies with cytoplasmic tails.

5. Integrin receptors

The receptors in the plasma membrane where components of the extracellular matrix adhere are now termed integrins. They are a family of transmembrane glycoproteins constructed of non-covalently linked heterodimers. The receptors to which, for example, fibronectin, vitronectin and platelet matrix components attach are all constructed to a similar plan, with the two subunits showing extensive amino acid homology (*Figure 3.12*).

The integrins serve as transmembrane links, connecting the extracellular matrix (*Table 2.2*) with cytoskeletal components (vinculin, talin, actin and tropomyosin). Integrins participate in cell–matrix and cell–cell adhesion processes involved in embryonic development, haemostasis, thrombosis, wound healing and immune defence mechanisms.

6. Cadherins

In addition to anatomically defined intercellular junctions described above, the specific association and attachment of cells to each other is mediated by membrane glycoproteins collectively called cadherins. These are receptor-like molecules that regulate Ca^{2+}-dependent intercellular adhesion, most thoroughly studied in embryonic cells and tissues, where they are believed to play a role in morphogenesis. However, they are present also in adult tissues, especially at areas of cell–cell attachment.

A number of cadherin subclasses have been identified. These include L-CAM, a liver adhesion molecule similar to uvomorulin, and cell-CAM, that mediates a number of adhesion-dependent processes in mammalian embryos and the formation of intercellular junctions. Other cadherins (e.g. N-CAM) are studied in neuronal tissues. As these 80–140-kd diverse molecules are being characterized, it emerges that they comprise a family of transmembrane glycoproteins with a high degree of identity in their amino acid sequences. As described in Section 4, they belong to the immunoglobulin superfamily. The molecular mechanisms underlying the cadherin-mediated interactions and their differential expression during development are being elucidated by analysis of the controlling genes.

7. Further reading

General
Dohlman,H.G., Caron,M.G. and Lefkowitz,R.J. (1987) *Biochemistry*, **26**, 2657–2663.
Huganir,R.L. and Greengard,P. (1987) *Trends Pharm. Sci.*, **8**, 472–477.

Insulin receptors
Goldfine,I.D. (1987) *Endocrinol. Revs.*, **8**, 235–255.
Maegawa,H., McClain,D.A., Freidenberg,G., Olefsky,J.M., Napier,M., Lipari,T., Dull,T.J., Lee,J. and Ullrich,A. (1988) *J. Biol. Chem.*, **263**, 8912–8917.

EGF receptor
Carpenter,G. (1987) *Annu. Rev. Biochem.*, **56**, 881–914.

LDL receptor
Goldstein,J.L., Brown,M.S, Anderson,R.G.W., Russell,D.W. and Schneider,W.J. (1985) *Annu. Rev. Cell Biol.*, **1**, 1–39.
Davis,C.G., Driel,I.R., Russell,D.W., Brown,M.S. and Goldstein,J.L. (1982) *J. Biol. Chem.*, **262**, 4075–4082.

Transferrin receptors
Stratford,M.M. and Cuatrecasas,P. (1985) *J. Membr. Biol.*, **88**, 205–215.

Acetylcholine receptor
Guy,H.R. and Hucho,F. (1987) *Trends Neurosci.*, **10**, 318–328.

Adrenergic receptor
Kobilka,B.K., Kobilka,T.S., Daniel,K., Regan,J.W., Caron,M.G. and Lefkowitz,R.J. (1988) *Science*, **240**, 1310–1316.

GABA receptor
Stephenson,F.A. (1988) *Biochem. J.*, **249**, 21–32.
Barnard,E.A., Darlison,M.G. and Seeberg,P. (1987) *Trends Neurosci.*, **10**, 502–509.

Neurotransmitter receptors
Strange,P.G. (1988) *Biochem. J.*, **249**, 309–318.

Rhodopsin and opsin receptors
Findlay,J.B.C. and Pappin,D.J.C. (1986) *Biochem. J.*, **238**, 625–642.

G-proteins
Neer,E.G. and Clapham,D.E. (1988) *Nature*, **333**, 129–134.
Stryer,L. and Bourne,H.R. (1986) *Annu. Rev. Cell Biol.*, **2**, 391–419.

Olfactory receptors
Anholt,R.R.H. (1987) *Trends Biochem. Sci.*, **12**, 58–62.

Immunoglobulin-type receptors
Williams,A.F. (1987) *Immunol. Today*, **8**, 298–303.
Williams,A.F. and Barclay,A.N. (1988) *Annu. Rev. Immunol.*, **6**, 381–408.

Integrin receptors
Ruoslahti,E. and Pierschbacher,M.D. (1987) *Science*, **238**, 491–497.
Hynes,R.O. (1987) *Cell*, **48**, 549–554.

HIV receptor
Sattentau,Q.J. and Weiss,R.A. (1988) *Cell*, **52**, 631–633.

Glucose transporters
Simpson,I.A. and Cushman,S.W. (1986) *Annu. Rev. Biochem.*, **55**, 1059–1089.

Phosphoinositol cycle
Berridge,M.J. (1987) *Annu. Rev. Biochem.*, **56**, 159–193.
Michell,R.H. (1988) *British Med. J.*, **295**, 1320–1323.

Oncogenes
Jove,R. and Hanafusa,H. (1987) *Annu. Rev. Cell Biol.*, **3**, 31–56.

Cell adhesion—cadherins
Takeichi,M. (1988) *Development*, **102**, 639–655.
Edelman,G.M. (1986) *Annu. Rev. Cell Biol.*, **2**, 81–116.

Protein kinases
Blackshear,P.J., Nairn,A.G. and Kuo,J.F. (1988) *FASEB J.*, **2**, 2957–2969.

Membrane biogenesis and trafficking

1. Introduction

The membrane proteins produced by ribosomes have to be transferred to the cellular sites where they function. A fundamental question arises as to how these proteins are accurately and efficiently targeted to their site of action. The same question, of course, applies to lipids synthesized on the smooth endoplasmic reticulum of eukaryotic cells.

In bacteria newly synthesized proteins are directed to either the inner or outer periplasmic membrane. In eukaryotic cells each membrane has a characteristic complement of proteins. All membrane proteins, with the exception of a few mitochondrial proteins, are coded for by nuclear genes and synthesized on ribosomes in the cytoplasm and these have to be transported to their final location where they carry out their functions.

The general exocytic and endocytic pathways of membrane protein traffic, as they occur in a polarized epithelial cell (e.g. a columnar epithelial cell of intestine or kidney) are shown in *Figure 4.1*. Proteins destined for the nucleus or mitochondria are transported to these organelles directly, whereas with those delivered to the Golgi, lysosomes, plasma membrane, the transfer of nascent polypeptides by ribosomes across the endoplasmic reticulum is involved. The proteins are then transferred onwards from the endoplasmic reticulum membrane probably by transitional carrier vesicles to the *cis* face of the Golgi apparatus where a number of modifications are introduced (see Section 2.2). The *trans* region of the Golgi apparatus directs the proteins to the cell surface and to the endocytic compartment and lysosomes. A constant cell surface area is maintained by a dynamic equilibrium of exocytosis and endocytosis.

2. Membrane biogenesis

2.1 Synthesis of membrane proteins

Mechanisms by which membrane proteins are inserted into the endoplasmic reticulum (co-translational protein transport), and for comparison, a soluble secretory protein, are shown in *Figure 4.2*. Two proteins, the signal recognition particle (a soluble particle containing six polypeptides and 7S RNA) on the

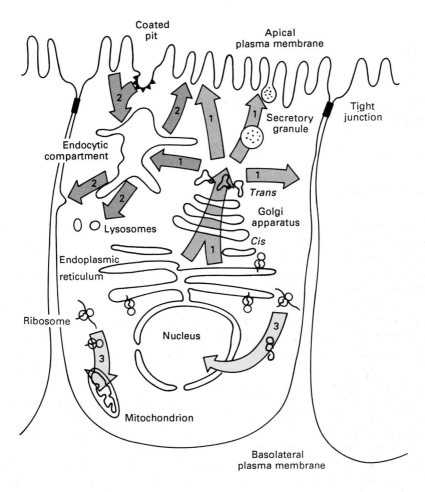

Figure 4.1.   Major membrane trafficking routes in an epithelial cell. **1**, The secretory pathways; **2**, endocytic pathways; **3**, nuclear and mitochondrial pathways.

ribosome and a 'docking protein', a 70-kd integral membrane protein, ensure the correct interaction of the ribosome with a cytoplasmic site on the endoplasmic reticular membrane. Translocation of the polypeptide chain through a membrane protein channel and cleavage of the signal peptide occur simultaneously during protein translation. Protein translocation requires ATP hydrolysis.

Signal peptides of a large number of membrane proteins have been sequenced. They are usually 16 – 26 residues long and contain a short (1 – 5 amino acid) basic region at the amino-terminus and a hydrophobic 'core' region that help anchor the protein in the membrane. This sequence is cleaved from the protein at the luminal side of the endoplasmic reticulum by a signal peptidase. However, many membrane proteins (e.g. asialoglycoprotein receptor, erythrocyte band 3 protein,

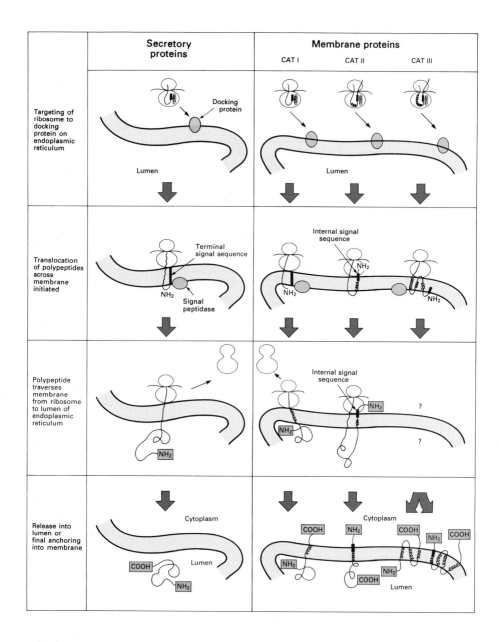

Figure 4.2.   Comparison of the membrane translocation of secretory proteins and three categories of membrane proteins. Secretory proteins are released into the lumen of the endoplasmic reticulum, but various stop-transfer and signal 'peptide' sequences combine to embed membrane proteins in the lipid bilayer. SRP, signal recognition particle.

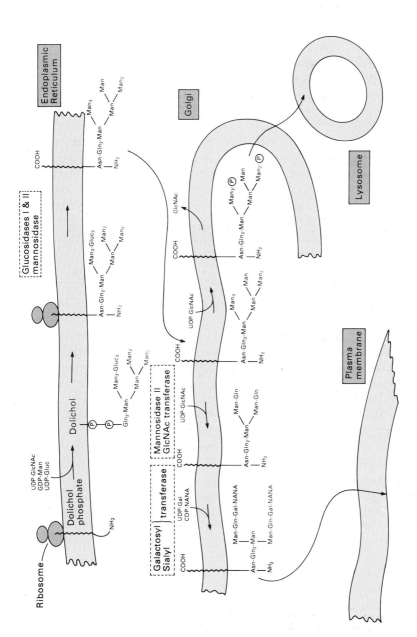

Figure 4.3.   Synthesis of membrane glycoproteins. Transfer from the endoplasmic reticulum to the Golgi involves 'transfer' vesicles. In the Golgi stacks, after trimming of mannose and glucose residues, the galactosyl and sialyl transferases convert the glycoprotein into its mature form. Distribution of membrane glycoproteins to various locations occurs from the *trans*-Golgi compartment. The figure depicts the route of an *N*-linked membrane glycoprotein. Gln, glucosamine; Man, mannose; Gluc, glucose; GlcNAc, *N*-acetylglucosamine. In the carbohydrate side chains Gln refers to *N*-acetylglucosamine.

Ca^{2+}-ATPase) are synthesized without incorporating a signal peptide at the amino-terminus. In these proteins, the internal hydrophobic membrane-spanning regions are thought to fulfil the functions of signal peptides. In bacteria, protein synthesis and transport are mainly electrochemical, post-translational events. Thus, many bacterial proteins are almost completely synthesized before they are inserted into the membrane.

In *Figure 4.2*, the synthesis is shown of three categories of integral membrane proteins that differ in topographical arrangement in the lipid bilayer. The biochemical properties of the categories of proteins are described elsewhere, but in brief, category I proteins include glycophorin, members of the immuno-gobulin superfamily, and the receptors for epidermal growth factor, insulin and low-density lipoproteins. Category II proteins, in contrast to those in category I, do not contain a cleavable signal-peptide sequence, but generally possess a hydrophobic amino acid sequence not far removed from the amino-terminus. Examples include the receptors for transferrin and asialoglycoprotein. Category III includes a large number of proteins that traverse the lipid bilayer several times, and examples include acetylcholine, GABA, adrenergic, muscarinic and rhodopsin receptors as well as various ion-activated ATPases. The mechanism by which these are threaded into the membrane is unknown.

2.2 Synthesis of membrane glycoconjugates

All eukaryotic cells synthesize membrane glycoproteins, and many cells also commit a part of their protein biosynthetic activity to the production of glyco-proteins destined for secretion. The synthesis of membrane glycoproteins occurs in the endoplasmic reticulum, Golgi apparatus and probably vesicles that shuttle between these membrane networks. The addition of sugars to proteins topo-graphically orientated in the various component membranes in the case of an *N*-linked glycoprotein is shown in *Figure 4.3*.

The oligosaccharide chain assembled on dolichol phosphate in the endoplasmic reticulum contains *N*-acetylglucosamine, mannose and glucose. Sugars, prior to their assembly into glycoproteins and glycolipids are transported into the membrane as sugar nucleotides for reaction with dolichol to form the 'core' polysaccharide containing *N*-acetylglucosamine, three glucose and nine mannose residues. Dolichol is a polyisoprenoid compound that is found in all eukaryotic cells. It is present mainly in endoplasmic reticulum membranes, where it is synthesized and in lower amounts in other membranes. Dolichol enhances vesicle fusion, suggesting a role in membrane trafficking, a property that is also governed by the nature of the attached fatty acid. The removal of the glucose residues by two glucosidases and the initial trimming of a single mannose residue on the 'core' polysaccharide by α-mannosidases have occurred before onward transfer of the nascent glycoprotein to the Golgi membranes.

In the Golgi apparatus, the final processing of glycoproteins occurs leading to the formation of the carbohydrate side chains that characterize individual glycoproteins. During this processing, up to six mannose residues in certain proteins are removed, involving the action of at least four mannosidases. A series of specific *N*-acetylglucosamine transferases, located mainly in the medial Golgi stacks, attach *N*-acetylglucosamine to the internal mannose core (see *Figure 4.2*). The galactosyl and sialyl transferase enzymes that then transfer galactose and sialic acid, respectively, to the glycoprotein are located mainly in the *trans* region of the Golgi apparatus. All these enzymes featuring in glycoprotein synthesis are located at the luminal aspect of the membrane. Many glycoproteins also contain fucose as a terminal sugar and this is attached in the Golgi by a fucosyl transferase.

The amino acid sequence in the protein that 'accepts' oligosaccharide side chains has been studied using various secreted and membrane glycoproteins. A consensus sequence of an internally-located tripeptide, Asn – X – Ser or Thr, in which X can be any amino acid except proline or aspartic acid has been identified with the sugars attached to the asparagine. However, not all polypeptides containing such a recognition sequence are glycosylated. Two major conditions governing glycosylation are the three-dimensional structure (conformation) of the glycoprotein and the location of the amino acid sequence on the luminal side of the membranes.

In addition to *N*-glycan glycoproteins described above, many glycoproteins are glycosylated on serine or threonine residues (*O*-glycans), and these proceed along a similar biosynthetic pathway. The amino acid sequence flanking the serine or threonine residues determines whether glycosylation proceeds. Indeed, many membrane proteins are not glycosylated as exemplified by the basic protein of myelin and the major protein of gap junctions (see Section 2 of Chapter 5). The biosynthetic routes followed by these proteins from the rough endoplasmic reticulum to the plasma membrane are not known.

A consequence of the general mechanism of the synthesis of membrane glyco-proteins shown in *Figure 4.3* is that sugar residues become topographically orientated towards the lumen of intracellular organelles and the external side of plasma membranes. However, glycoproteins of nuclear membranes constitute a different class with a unique pattern of glycosylation for they lack sialic acid, fucose and *N*-acetylgalactosamine. These glycoproteins are often associated with nuclear pores and contain a single *N*-acetylglucosamine residue. They are located at the cytoplasmic and nucleoplasmic aspect of the nuclear membranes. Other classes of *O*- and *N*-acetylglucosamine-linked proteins orientated to the cytoplasmic side of the endoplasmic reticulum have also been identified recently.

Many membrane glycoproteins are covalently modified by sulphation on tyrosine residues, a process occurring in the Golgi apparatus and catalysed by a tyrosyl protein sulphotransferase. Sulphation involves the transfer to the protein of sulphate from the donor 3′-phosphoadenosine 5′-phosphosulphate. Many sulphated glycoproteins are secreted or form part of the extracellular matrix (e.g. heparan sulphate and fibronectin). Other components of the extracellular matrix are glycosylated first and the carbohydrate residues are then sulphated in the Golgi apparatus.

2.3 Synthesis of membrane lipids

Membrane lipids are synthesized in the endoplasmic reticulum where all the

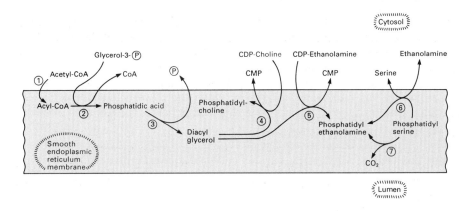

Figure 4.4. Phospholipid biosynthesis in the endoplasmic reticulum. Enzymes involved are: 1, fatty acid synthesis enzymes; 2, glycerol-3-phosphate acyl transferase and lysophosphatidic acid acyl transferase; 3, phosphatidic acid phosphatase; 4, diacylglycerolcholine phosphotransferase; 5, diacylglycerolethanolamine phosphotransferase; 6, phosphatidylethanolamine serine transferase; and 7, phosphatidylserine decarboxylase.

participating enzymes are located. Phosphatidylcholine is synthesized from activated fatty acids (fatty acid-coenzyme A) and glycerol phosphate to produce phosphatidic acid that is then converted to form the phospholipids shown in *Figure 4.4.* Lipid synthesis is a highly asymmetrical process for all the enzymes involved have their active sites located at the cytosolic aspect of the endoplasmic reticular membrane. As the lipid bilayer expands, mechanisms, probably enzymic, exist to allow individual phospholipids to 'flip-flop' between the two halves of the bilayer resulting in the creation of the phospholipid asymmetry observed in the various membrane systems.

In the synthesis of sphingomyelin, sphingamine (formed by condensation of serine with palmityl-CoA) is acylated to form ceramide. The phosphorylcholine head group is transferred to ceramide yielding sphingomyelin. The synthesis of phosphoinositides is shown in Section 2.3 of Chapter 3, and the glycosylation of ceramides by glycosyltransferases located in the Golgi apparatus is described in Section 2.2.

Cholesterol is also synthesized predominantly in the endoplasmic reticulum where the key regulatory enzyme is β-(OH)-β-methylglutaryl-CoA reductase (HMG-CoA reductase). The enzyme converts HMG-CoA to mevalonate, which is the precursor of squalene that ultimately cyclizes to form cholesterol. Cholesterol synthesis is complex and self-regulated. Cholesterol can suppress the transcription of the gene coding for HMG-CoA reductase. Clearly, drugs acting on this key enzyme can potentially lower the rate of cholesterol biosynthesis, and thus, indirectly, diminish the levels of circulating low-density lipoproteins.

Glycerol phosphate is also synthesized in mitochondrial membranes and by chloroplasts. Cardiolipin is formed by condensation of phosphatidylglycerol and CDP-diacylglycerol, a process occurring in the inner mitochondrial membrane where this lipid is exclusively located.

Lipids are also glycosylated to yield glycolipids that are found at high concentrations at the plasma membrane. The most thoroughly studied glycolipids are those in erythrocyte membranes that constitute the blood group substances. Glycolipids are synthesized in the endoplasmic reticulum and the Golgi apparatus. The sugars are attached in specific sequences by similar mechanisms and enzymes to those elaborating glycoproteins, involving the transfer of sugar nucleotides to the hydroxyl group of ceramide, followed by sequential additions of further sugars.

3. Exocytosis, secretory vesicles and membrane fusion

Exocytosis involves the fusion of the membranes of secretory vacuoles or granules with the inner face of the plasma membrane (*Figure 4.5*). It is an event that is accentuated in secretory tissues, mast cells and in the neuroendocrine cells. During the fusion of the intracellular secretory granules with the plasma membrane, the granules are observed to swell. It is unclear whether this swelling precedes membrane fusion or follows it. The molecular rearrangements in the membranes undergoing fusion are as follows.

First, an ion-conducting channel forms early in the fusion process. The pore, of diameter 2 nm, forms in less than 0.1 msec. The secretory vesicle has to penetrate through a passage cleared in the cytoskeleton before it can approach the inner face of the plasma membrane. In the nervous system, synapsin 1, a

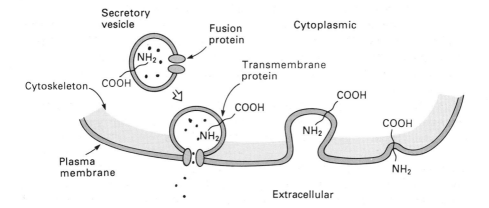

Figure 4.5.   Fusion of a secretory vesicle with the plasma membrane. Secretory vesicles approaching the cell surface have to penetrate the cytoskeleton underlying the plasma membrane. Specific proteins facilitate fusion of secretory vesicles with the plasma membrane.

neuron-specific phosphoprotein associated with synaptic vesicles is believed to help dock the vesicle as it interacts with components of the cytoskeleton—especially actin and villin. In adrenal chromaffin granules, fodrin, a non-erythroid form of spectrin, is implicated in the exocytic event. Several other proteins are also implicated in facilitating exocytosis including caldesmon (a calmodulin-binding F actin regulatory protein) and gelsolin. These and probably other proteins that regulate calcium levels facilitate the penetration of the secretory vacuole through the microfilament barrier located underneath the plasma membrane, a necessary prelude to the act of exocytosis (*Figure 4.5*).

In delineating the movement of secretory vesicles from the Golgi apparatus through the cytoplasm for fusion with the plasma membrane, it is useful to distinguish between the mechanisms of secretion operational in various cell types. In endocrine cells, for example, proteins to be secreted are first concentrated as much as 200-fold during transfer from the Golgi cisternae to the secretory vesicle. Secretion is highly regulated and the vesicles are prevented from fusing with the plasma membrane until the level of a cytoplasmic messenger, usually Ca^{2+}, is altered. In other cells, for example hepatocytes, fibroblasts, muscle and yeast cells, secretion is a constitutive (constant) event. There is no external stimulus and these cells alter the rate of exocytosis by varying the rate of protein synthesis. In polarized cells, there is additional complexity, for secretion is directional, requiring targeting to a specific domain of the plasma membrane. For example, in neurons, membrane vesicles departing from the Golgi apparatus are directed to move down the axon to the nerve terminal; microtubules are associated with this movement. It is also thought likely that similar mechanisms occur in other cells, involving a key role for microtubules in directing secretory vesicles in an orderly fashion to the plasma membrane.

4. Sorting and targeting of membrane proteins

A multiplicity of proteins are produced in the endoplasmic reticulum. The task of identifying the topogenic signals that retain those proteins that function in the endoplasmic reticulum while allowing others to migrate to the other membrane systems is a major goal. Information on the precise targeting signals for the individual proteins is currently sparse but progress has been made in identifying the first of the myriad amino acid sequences and other 'address' signals underlying the sorting of proteins.

4.1 Signals for retention in the endoplasmic reticulum

One general view held on the sorting and dispatch of newly synthesized membrane proteins from the endoplasmic reticulum is that no specific signal (i.e. amino acid sequence) is necessary for these to flow rapidly, and in bulk, through the Golgi apparatus and onwards to the plasma membrane, lysosomes, etc. However, the presence and identification of a discrete signal allows the retention of the permanent residents within the endoplasmic reticulum network. For example, the sequence Lys – Asp – Glu – Leu has been identified as necessary to retain three non-membrane proteins in the endoplasmic reticulum.

4.2 Targeting to mitochondria

Mitochondrial proteins, of which there are several hundred, are synthesized by two genetic systems. The mitochondrial genome codes for about 10 proteins of the inner membrane (see *Figure 5.6*) but the vast majority of the proteins are coded for by nuclear genes, synthesized mainly by free ribosomes in the cytosol and then imported across the mitochondrial outer membrane.

The trafficking of proteins from the cytosol into mitochondria requires two types of sorting signals—those for directing the protein towards the mitochondria and those for targeting to either the inner or outer mitochondrial membranes and the matrix spaces. Proteins directed to mitochondria first bind specifically to the cytoplasmic face of the outer mitochondrial membrane at sites called 'import receptors'. Many proteins entering mitochondria contain a signal-type amino acid pre-sequence that can vary in size between 0.5 kd and 10 kd. This pre-sequence is cleaved from the protein as it is positioned in the mitochondrial membrane. Analysis of the amino acid sequences in a variety of cleaved pre-sequences indicates that they are rich in arginine and lysine residues, usually separated from one another by uncharged amino acids such as serine and threonine; acidic amino acid sequences are generally absent. This suggests that it is the overall amino acid sequence that governs mitochondrial targeting. The mitochondrial precursor proteins traverse from the outer to the inner membranes at specific regions of contact.

4.3 Targeting to plasma membrane domains

The differentiation of the plasma membrane, especially in epithelial cells, into basolateral and apical domains poses the question as to how this arises—is the domain structure of cell surfaces established by random insertion of proteins followed by a selective lateral redistribution in the membrane or by direct insertion into specific domains? To what extent do tight junctions (see Section 5 of Chapter 1) regulate the regionalization of proteins and lipids between basolateral and apical domains? Much of our knowledge of the sorting of membrane proteins arises from studies using viral glycoproteins as markers. Certain enveloped RNA viruses (e.g. influenza virus) exit from cells by budding at the apical plasma membrane domain whereas others (e.g. vesicular stomatitis virus) bud at the basolateral domain. Studies using cultured cells infected with viruses show that sorting of glycoproteins may occur at the *trans* region of the Golgi apparatus but in 'working' epithelia (e.g. liver, kidney) in which there is constant ligand-induced endocytic trafficking from the basolateral to the apical poles of the cell the situation is more complex (see Section 5.3).

4.4 Targeting to the nucleus

Specific amino acid sequences that direct proteins to the nucleus are being derived and one such nuclear 'import' or karyophilic signal is Pro – Lys –

Lys – Lys – Arg – Lys – Val. Entry into the nucleus of large proteins consists of two steps—first the protein binds to the nuclear pore followed by an energy-dependent translocation through the pore.

4.5 Targeting to lysosomes and peroxisomes

The signal targeting of certain newly synthesized glycoproteins to lysosomes involves the attachment in the Golgi apparatus of a phosphomannose residue (see Section 2.2). Receptors recognizing phosphomannosyl glycoproteins are widespread in animal cells, and they shuttle enzymes, for example acid hydrolases, to their functional locus in lysosomes. In a pre-lysosomal compartment, the low pH releases the enzymes allowing the receptors to shuttle back to the Golgi apparatus. A dipeptide (Ser, Lys or His, Leu) at the carboxyl-terminus directs enzymes to peroxisomes.

4.6 Targeting to chloroplasts

Chloroplasts contain three membrane systems requiring cytoplasmically synthesized proteins to be directed to their correct functional location (see *Figure 5.7*). Thus, a protein destined for the thylakoid membrane has to traverse two membranes and the stroma.

Proteins targeted to chloroplasts are synthesized as higher-molecular-weight precursors. The additional amino acid sequence, called a 'transit peptide', is located at the amino-terminus, and is rich in hydroxylated, basic and small hydrophobic amino acids. Binding to the chloroplast surface may be mediated by a receptor protein, and translocation across the chloroplast double membrane is ATP-dependent, although the exact mechanism by which its hydrolysis is connected to protein translocation is unknown. Movement across the membrane may involve threading the unfolded protein or implicate translocation proteins. Clearly, there are many points of similarity between targeting to mitochondria and chloroplasts. Little is understood of the criteria for targeting proteins onwards to thylakoid membranes and their subsequent assembly into oligomers. Indeed, many of the thylakoid membrane proteins, including some involved in photosynthesis, are synthesized by ribosomes bound to the thylakoid membrane and inserted co-translationally into the membrane.

5. The plasma membrane and endocytosis

One of the major functions of the plasma membrane is the controlled uptake of a variety of materials from the environment. Pinocytosis involves uptake of fluid material, whereas phagocytosis involves particulate material. Both processes involve invagination of patches of the plasma membrane and encirclement of the material to be endocytosed (*Figure 4.6*).

Endocytotic processes are dynamic and energy consuming. Extreme examples are provided by tissue culture cells that internalize 30% of the total surface area each hour, and by macrophages that internalize by pinocytosis up to 186% of

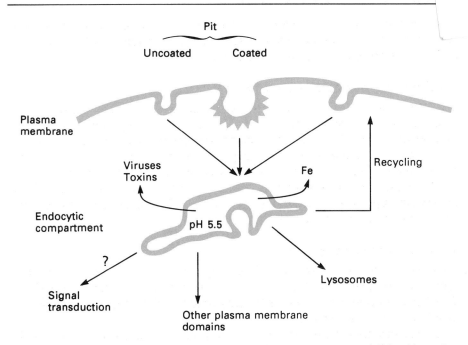

Figure 4.6. A generalized view of endocytosis. Arrows show routes followed by endocytosed ligands. Iron (Fe) is internalized, bound to transferrin, then released. Hormones and growth factors may trigger signal transduction, e.g. protein phosphorylation.

their surface area each hour. The fact that the surface area of these cells remains fairly constant indicates that internalized membrane components are rapidly recycled back to the cell surface.

Two types of pinocytotic endocytosis are now recognized. Fluid phase endocytosis is non-selective, with the solutes being internalized across unspecialized surface areas in proportion to their concentration in the external medium. Adsorptive endocytosis, in contrast, is highly selective and involves specialized coated regions of the plasma membrane. Importantly, the selectivity and concentrating power is provided by specific interactions of the ligands with cell surface receptors, hence the term receptor-mediated endocytosis.

5.1 Coated pits

The major consequence of the specific binding of circulating ligands, both large (e.g. viruses, lipoproteins) and small (e.g. polypeptide hormones, many drugs) to cell surface membrane receptors is a rapid transfer into a membrane-enclosed system termed the endocytic compartment. Many of the ligand–receptor complexes are first concentrated, by lateral movements in the plane of the plasma membrane, into morphologically specialized domains called coated pits. The characteristic underlying coat is composed of a protein termed clathrin, constructed of heavy (180 kd) and two light (35–40 kd) polypeptide chains, as well as a number of assembly polypeptides. Some membrane proteins, such as the low-density lipoprotein receptor after binding its ligand (Section 3.2 of

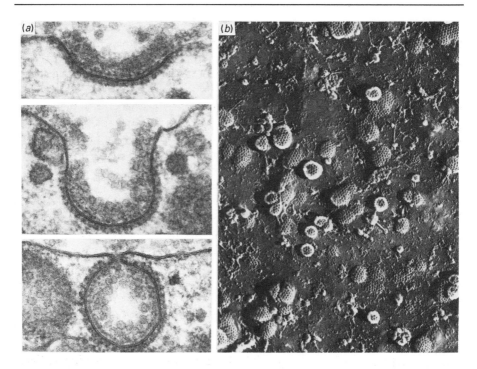

Figure 4.7. (a) Electron micrograph showing formation of a coated pit. This sequence shows a hen oocyte interiorizing lipoprotein particles [courtesy of M.M.Perry; from *J. Cell Sci.* (1979), **39**, 257]. (b) Electron micrograph of the inner surface of an hepatocyte plasma membrane showing coated pits. A shadow replica of the cytoplasmic aspect of the substratum-adherent plasma membrane was made (courtesy of M.Nermut).

Chapter 3), are concentrated into coated pits about 100-fold over the surrounding plasma membrane; others, such as adenylate cyclase, are largely excluded.

Coated pits thus function as molecular filters controlling the uptake by cells of circulating ligands that have become attached to receptors. They are also responsible for invagination of the plasma membrane to form endocytic vesicles (*Figure 4.7*). The ligand–receptor complexes are then transferred within minutes from the coated pits into the endocytic compartment involving an enzyme-catalysed uncoating of the internalized membrane.

5.2 The endocytic compartment

Ligand–receptor complexes are transferred from coated pits in the plasma membrane into an extensive network of anastomizing vesico-tubular membranes called the endocytic compartment. Endocytosis is a continuous activity in most cells; many receptors are internalized and then returned to the cell surface in the absence of bound ligands. As shown in *Figure 4.6*, current research indicates that transfer of ligands from the plasma membrane may also occur from uncoated membrane regions, since many cells continue to endocytose ligands when formation of coated pits is blocked. The endocytic compartment interacts functionally with lysosomes, the Golgi apparatus and the various regions of the plasma membrane and it is the prime regulator of inward membrane traffic.

The membranes of the endocytic compartment carry out many functions. The low pH generated in the lumen of the compartment by a proton ATPase results in the dissociation of many of the internalized ligand – receptor complexes thus permitting the ligands and the receptors to be processed independently; this property accounts for the acronym CURL (compartment for uncoupling receptor and ligand) being applied to these membranes. Certain receptors, for example those at which asialoglycoproteins bind, have been shown to become concentrated at membranes at the rim of the compartment prior to their return to the plasma membrane (the retroendocytic vesicular pathway). The fate of ligands in the endocytic compartment can vary; many are transferred to lysosomes where degradation occurs, but other ligands, for example Fe^{2+}, viruses, and toxins, move across the membrane into the cytoplasm. The role of the endocytic compartment in the sorting of internalized membrane constituents and attached ligands attains an additional degree of complexity in polarized cells, where some receptors and ligands are dispatched onwards towards the opposite pole of the cell. This route, evident in epithelia, is called transcytosis (see Section 5.3).

New functions are being attributed to the endocytic compartment. In view of the presence in the compartment of polypeptide hormones and growth factors attached to their receptors it is possible that transmembrane signalling, initiated at the plasma membrane, continues. Components of the extracellular matrix, for example fibronectin and proteoglycans, are also found associated with endocytic membranes suggesting a role in their modulation and processing. Certain ligands, e.g. insulin and epidermal growth factor also undergo proteolytic trimming in the endocytic compartment. This processing is distinct from the extensive breakdown that occurs when the ligands are transferred to lysosomes.

Two morphological regions of the endocytic compartment are distinguished— the membrane vesicles proximal to the plasma membrane where various ligands are first detected after uptake, and multivesicular membrane bodies in deeper regions found, in some cells, close to the nuclear membrane. At present it is unclear whether membrane components of the endocytic compartment change gradually as they move deeper into the cell, or whether two biochemically distinct membrane networks exist which are connected by a population of carrier vesicles. Microtubules may be involved in regulating membrane movement in the deeper regions of the compartment.

5.2.1 Endocytosis of viruses

Entry of enveloped viruses into cells involves binding to cell surface receptors, followed, in many instances, by endocytosis. After transfer of viruses into the endocytic compartment, the low pH results in conformational changes in their structure resulting in fusion with the endocytic membrane and extrusion of the viral nucleocapsid into the cytoplasm.

Viruses combine with a variety of receptors at the cell surface, and the nature

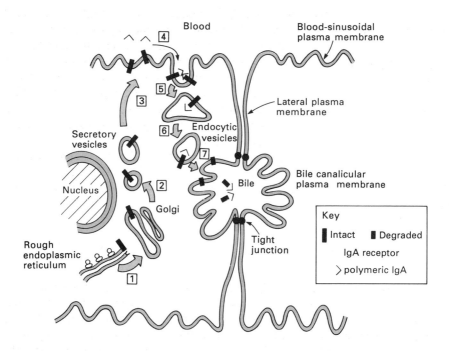

Figure 4.8.   Itinerary of the IgA receptor in hepatocytes. The following segments are shown. 1, Endoplasmic reticulum to the Golgi apparatus; 2, Golgi apparatus to secretory vesicles; 3, secretory vesicles to the blood sinusoidal plasma membrane; 4, concentration of receptor with bound polymeric immunoglobulin into a coated pit; 5, endocytosis of receptor – ligand complex into peripheral endocytic compartment; 6, transport of receptor – ligand complex in endocytic vacuoles towards biliary pole of hepatocyte; and 7, transfer across bile canalicular plasma membrane and release of degraded receptor and polymeric immunoglobulin A into bile.

of these receptors is known in many cases. For example, adenovirus and human cytomegalovirus bind to the Class I (HLA) histocompatibility antigen and the HIV (human immunodeficiency virus) binds to the CD4 receptor on the plasma membrane of T lymphocytes. There is much variation in the specificity of viruses for cell surface receptors, with many viruses binding to a range of sialoglyco-proteins and glycolipids.

5.3 Transcytosis

Many epithelial cells transport ligands across the cell interior from one extra-cellular compartment to another, a function that involves the sorting capacity of the membranes comprising the endocytic compartment. For example, polymeric IgA is specifically taken up from blood by a variety of glandular epithelial cells and transported across the cell and released into external secretions such as milk, bile and intestinal fluid which form the first immuno-logical defence against infection.

The receptor which binds polymeric IgA contains an ectoplasmic portion called secretory component. As shown in *Figure 4.8*, in hepatocytes the receptor is synthesized in the rough endoplasmic reticulum and is transported through the Golgi apparatus to the baso-lateral region of the cell surface. Polymeric IgA circulating in the blood binds to the receptor and the receptor – IgA complex is then internalized into the endocytic compartment and transported to the apical surface where both components are released outside the cell. In the hepatocyte, as shown in *Figure 4.8*, the receptor, before its discharge by membrane shedding into bile undergoes partial proteolysis at its extracellular domain.

The polymeric IgA receptor is transferred from the rough endoplasmic reticulum to the sinusoidal plasma membrane in 30 – 60 min. Transcytosis from the sinusoidal to the bile canalicular plasma membrane via endocytic vesicles takes about 10 – 30 min.

6. Lipid transport pathways

Most of the enzymes involved in lipid biosynthesis are resident in the endoplasmic reticulum, requiring lipids, like proteins, to be transported and targeted to various parts of the cell. However, much less is known of the intracellular trafficking of lipids. The use of fluorescently labelled phospholipids as markers is beginning to shed some light on the underlying mechanisms. These studies reinforce the view that lipids do not move in a random fashion in cells, but travel rapidly and specifically to and between intracellular organelles. The mechanism of lipid transport out of the endoplasmic reticulum and between various organelles is not known although it is generally thought that vesicular traffic accounts for the discrete transfer of lipids and proteins between various organelles. Fusion of vesicles with other membranes, a process dictated by specific protein 'acceptors', can account for the characteristic compositions of the membrane systems and organelles whilst allowing the discrete delivery of lipids and proteins, that is, vesicles have a built-in targeting capacity.

7. Membrane turnover

Cell membranes are dynamic entities, with their lipid and protein components constantly undergoing synthesis and degradation. Lipids and proteins turnover at different rates and there is a wide variation in the half-lives of individual proteins—from a few hours to days. Lipids are degraded by specific phospholipases hydrolysing ester bonds in phospholipids. A wide range of proteases feature in the relatively slower protein degradation. Although proteases are present at high concentrations in lysosomes, their ubiquity in membranes suggests that specific modification, processing and degradation of membrane proteins occurs widely in the cell. With glycoproteins, the half-life of the carbohydrate moieties is considerably less than that of the polypeptide backbone. For example, studies of a liver plasma membrane glycoprotein showed that the protein had a half-life of 63 h, whereas the mannose half-life was 41 h, and that of galactose 23 h.

Many animal cells release discrete plasma membrane fragments into the surrounding environment. This membrane shedding (exfoliation) is more extensive in cells that have difficulties in balancing secretory and endocytotic activities (e.g. malignant cells). The release of membrane fragments is governed by the underlying cytoskeleton.

8. Further reading

Protein assembly
Racker,E. (1987) *Science*, **235**, 959–961.

Protein translation
Walter,P. and Lingappa,V.R. (1987) *Annu. Rev. Cell Biol.*, **2**, 499–516.

Glycosylation; topology
Lennarz,W.J. (1987) *Biochemistry*, **26**, 7205–7210.
Verbert,A., Cacan,R. and Cechelli,R. (1987) *Biochimie*, **69**, 91–99.

Secretory pathways
Kelly,R.B. (1985) *Science*, **230**, 25–32.
Bourne,H.R. (1988) *Cell*, **53**, 669–671.

Golgi apparatus and sorting
Griffiths,G. and Simons,K. (1986) *Science*, **234**, 438–443.
Pfeffer,S.R. and Rothman,J.E. (1987) *Annu. Rev. Biochem.*, **56**, 829–852.

Endocytosis and membrane traffic
Yamashiro,D.J. and Maxfield,F.R. (1988) *Trends Biochem. Sci.*, **9**, 190–193.
Tartakoff,A.M. (1987) *The Secretory and Endocytic Pathways*. Wiley-Interscience, New York.
Morré,D.J, Howell,K.E., Cook,G.M.W. and Evans,W.H. (eds) (1988) *Cell-free Analysis of Membrane Traffic. Proceedings of Clinical and Biological Research*, Vol. 270, A.R.Liss, New York.
Geuze,H.J., Slot,J.W. and Schwartz,A.L. (1987) *J. Cell Biol.*, **104**, 1715–1723.

Viral entry and membrane fusion
Wiley,D.C. and Skehel,J.J. (1987) *Annu. Rev. Biochem.*, **56**, 365–394.

Coated pits
Pearse,B.M.F. and Crowther,R.A. (1987) *Annu. Rev. Biophys. Biophys. Chem.*, **16**, 49–68.
Moore,M.S., Mahaffey,D.J., Brodsky,F.M. and Anderson,R.G.W. (1987) *Science*, **236**, 558–563.

Transcytosis
Solari,R. and Kraehenbuhl,J.-P. (1985) *Immunol. Today*, **6**, 17–20.

Nuclear targeting
Dingwall,C. and Laskey,R.A. (1986) *Annu. Rev. Cell Biol.*, **2**, 367–390.
Newmeyer,D.D. and Forbes,D.J. (1988) *Cell*, **52**, 641–653.

Topogenic sequences in protein sorting
Robinson,A. and Austen,B. (1987) *Biochem. J.*, **246**, 249–261.

Protein sorting
Rothman,J.E. (1987) *Cell*, **50**, 521–522.
Pfeffer,S.R. (1988) *J. Membrane Biol.*, **103**, 7–16.

Lipid sorting
Van Meer,G. and Simons,K. (1988) *J. Cell. Biochem.*, **36**, 51–58.
Pagano,R.E. and Sleight,R.G. (1985) *Science*, **229**, 1051–1057.
Simons,K. and Van Meer,G. (1988) *Biochemistry*, **27**, 6197–6202.

Membrane transport and bioenergetics

1. Ion carriers

Membranes are intimately involved in controlling changes in ion movement between the cell and the external environment as well as between various intracellular compartments. For example, there is a 10 000-fold difference in Ca^{2+} concentration between the inside of the cell and its environment. Ion fluxes across membranes are closely linked to cell growth and proliferation. Various enzyme and exchange mechanisms regulate the alkalinity or acidity, the concentrations of Na^+, K^+, Ca^{2+} and Cl^- and other anions of organelles or membrane-lined vacuolar compartments. Such gradients are established at the expense of metabolic energy, e.g. ATP hydrolysis. Ionic gradients are used to transfer solutes across membranes (e.g. amino acids, sugars) into the cell by symport carriers and protons out of the cell by antiport carriers. The outward K^+ gradient generated across the plasma membrane is a major determinant of the inside negative transmembrane potential (~ 40 mV) of cells. In epithelial tissues, the polarized distribution of enzymes and ion carriers provides the driving force for movement of ions and molecules across the cell interior. Ion movements, especially proton transport, underlie the synthesis of ATP in mitochondria, photosynthesis in plants and substrate uptake by bacteria.

2. Gap junctions

Cells 'talk' or communicate to each other in a variety of ways. For example, cells release hormones and growth factors into the blood and these influence the behaviour of other cells after binding to receptors at the cell surface membrane (Chapter 3). Cells also communicate directly across gap junctions. These are channels present in discrete areas of the plasma membrane that allow direct continuity between the cytoplasms of adjacent cells. The channel ($\sim 1.6-2.0$ nm in diameter) is large compared with other ion channels. Gap junctions permit ions and metabolites of up to about 1 kd to pass through the channel in both directions. They have now been detected in most animals, from hydra to mammals, and they are present in all tissues and organs with the exception of striated muscle. Gap junctions display a characteristic morphology,

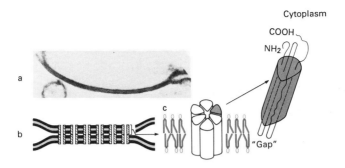

Figure 5.1. Morphology and topography of the major gap junction protein. **(a)** Electron micrograph of a liver gap junction composed of two converging areas of the plasma membrane. **(b)** Diagrammatic representation of **a**. **(c)** Model of connexon in junctions, two connexons are orientated end-to-end allowing formation of a channel extending across the two plasma membranes. Each connexon may consist of six 32-kd polypeptides. The possible orientation of the polypeptide in liver is indicated in the blow-up of a single subunit.

especially when viewed by freeze-fracture electron microscopy which reveals an ordered pattern of intramembranous particles, and by negative staining of isolated junctions in which crystalline hexagonal arrays are evident. A minor variation in morphology is evident in arthropods, where the intramembranous particles are of a larger size.

Figure 5.1 shows the structure of the gap junction of an hepatocyte. It is composed of two sets of connexons contributed by the two plasma membranes joined in mirror symmetry. Each connexon, in turn, is constructed of six integral membrane proteins, and a low-resolution model of the major 32-kd protein is shown. The component proteins of gap junctions—termed connexons—have been isolated and the amino acid sequences determined indirectly from the cDNA in a number of tissues. Although connexons vary in size, mainly according to the length of the carboxyl-terminal tail located at the cytoplasmic aspect of the junction, they are built to a common plan, which explains the morphological similarity of gap junctions between species and tissues.

The gap junction channels close rapidly when intracellular Ca^{2+} concentrations increase or pH drops (e.g. during cellular injury) but the underlying changes are not fully understood. Gap junctions form rapidly between co-operating cells after contact is made, indicating that the junction is assembled from pre-existing subunits that can move laterally in the plane of the membrane.

3. Cation carriers

Cation channels exist in open or closed conformations. When open, ions flow through the channel to produce an electric current. For example, sodium channels controlling the transient sodium current during an action potential are

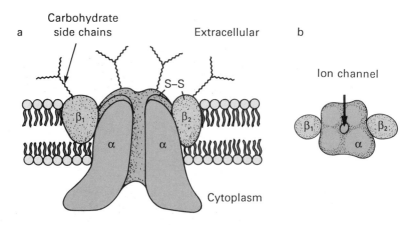

Figure 5.2. Model of the sodium channel. **(a)** Topography of polypeptide subunits in the membrane. **(b)** Surface view of channel.

characteristic of most nerve and muscle cell membranes. Many drugs, such as local anaesthetics and anti-dysrhythmics block these channels and so interrupt transiently the flow of current through them. These voltage-gated channels open as a result of a change in the potential gradient across the membrane. A further class are the receptor-operated channels that open as a result of the binding of neurotransmitter molecules to sites on their extracellular aspect (see Section 2.4 of Chapter 3).

The most intensively studied cation channels are those allowing passage of Na^+ and Ca^{2+}. A low resolution model of the voltage-gated sodium channel is shown in *Figure 5.2*. The sodium channel in rat brain is composed of a 260-kd α subunit and two β subunits, 36 kd and 33 kd. All the subunits are glycosylated and there is some slight variation in the size of the β subunits in various tissues. Neurotoxins with well described physiological effects bind to sodium channels with high affinity and specificity and modify their properties. These include tetrodoxin, and scorpion, sea anemone, snake and bacterial toxins. Indeed, these toxins have helped in the identification and purification by affinity chromatography of the sodium channel proteins.

Calcium channels are also present in plasma membranes and have been investigated in muscle and neuronal cells where they are influenced by drugs, for example verapamil and other Ca^{2+} channel antagonists. When channels open, they allow Ca^{2+} ions to move down their electrochemical gradient across the plasma membrane into the cytoplasm. The entry of Ca^{2+} into the cell provokes specific responses, e.g. the fusion of synaptic vesicles with the pre-synaptic membrane in nerve cells, hormone release by endocrine tissue and activation of contractile elements in muscle. Controlling the response of cells to intracellular Ca^{2+} levels are various Ca^{2+}-binding proteins (e.g. calmodulin, calpactin) that may bind to the Ca^{2+}-ATPase pumping Ca^{2+} out of the cell (*Figure 5.3*).

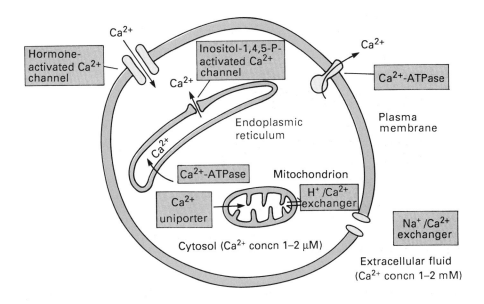

Figure 5.3. Various calcium translocation mechanisms—energized and non-energized in an animal cell.

4. Anion carriers

Anions cross membranes by three general mechanisms as shown in *Figure 5.4*. The first is a passive electrodiffusion and this is usually unidirectional; such anion 'leaks' are extremely small in normal cells. The second pathway involves electro-neutral exchange of anions for example Cl^- for HCO_3^-; this is called anion antiport. The third mode of anion transport involves the co-transport of anions and cations in a specific manner; for example the entry of Na^+ into cells is accompanied by Cl^- and these are called symport mechanisms. The corollary involves the subsequent removal of Na^+ by the Na^+/K^+-ATPase during which there is an inward transport of K^+.

Anion-exchange pathways have been studied in greatest detail in erythrocytes where they play a key role in CO_2 transport. For example, CO_2 diffusing into cells is converted into H^+ and HCO_3^- by carbonic anhydrase. The HCO_3^- is then exchanged for plasma Cl^- whereas the H^+ is buffered by haemoglobin. Importantly, the exchange of HCO_3^- for Cl^- across the cell membrane, thereby removing a product of the carbonic anhydrase reaction, allows CO_2 to be stored as plasma bicarbonate and greatly increases the CO_2-carrying capacity of blood. Specific channels in the erythrocyte membrane facilitate anion-exchange, which is an extremely fast process taking $40-50$ msec. The anion carrier is present in protein 3, the major transmembrane protein in the erythrocyte, accounting

Table 5.1. Comparison of proton ATPases

Properties	Vacuolar	Eubacterial/ mitochondrial	Plasma membrane
Size (kd)	200 – 500	450 – 550	100
Subunits	3 – 8	8 – 12	1
Electrogenic[a]	Yes	Yes	Yes
Inhibitors	N-ethylmaleimide Nitrate	Oligomycin Azide	Vanadate
Ion requirement	Cl^1	–	K^+
Phosphorylated intermediate	No	No	Yes

[a]An electrogenic process is one in which the proton translocation is not directly coupled to the movement of other ions. An electroneutral process is one in which pumping involves exchange of other ions.

for 25 – 30% of the total membrane protein (*Figure 2.6*). There are about one million molecules of protein 3 per cell. In addition to protein 3 studied in erythrocytes, a general class of non-erythroid anion channels with a similar structure exists in epithelial tissues where they fulfil anion-exchange processes.

5. Membrane ATPases

5.1 Proton ATPases

Various ATPases control the proton concentrations in cell compartments and organelles. The proton-motive force plays a major role in mitochondrial and photosynthetic processes. Proton ATPases are classified into three main families (*Table 5.1*).

   Vacuolar proton pumps are present in organelles connected to the vacuolar system of eukaryotic cells and they are also found in bacteria, fungi and plant vacuoles. They function in the controlled pumping of protons into organelles thereby acidifying the inside of the organelles and membrane vesicles and generating a proton-motive force that can be utilized for a secondary uptake process. In lysosomes, the proton pump lowers the pH in the lumen to about pH 5 which is the optimal pH of most lysosomal enzymes. Proton pumps also acidify the interiors of the endocytic compartment and the *trans* region of the Golgi apparatus (see Chapter 4). Movement of exocytic (secretory) vesicles is also dependent on controlled acidification. The proton ATPase has been studied in great detail in adrenal chromaffin granules where it generates the proton-motive force responsible for the uptake of catecholamines.

   Proton pumps present in bacteria, mitochondria and chloroplasts harness the proton-motive force generated by mitochondrial and photosynthetic electron transport for the phosphorylation of ADP (Section 5.6). In bacteria ATP formed by fermentation is used to generate a proton-motive force that is utilized by secondary uptake systems. The bacterial/mitochondrial proton pumps differ in

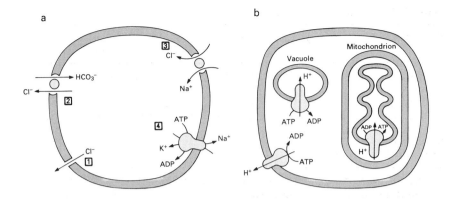

Figure 5.4. Ion-transport pathways in animal cells. (a) Anion-transport pathways: 1, passive electrodiffusion; 2, anion antiport; 3, symport mechanisms; 4, active transport involving Na^+/K^+-ATPase. (b) Location of proton pumps.

evolutionary origin from the other pumps described and their structure supports the view that mitochondria and chloroplasts have evolved from prokaryotes.

Plasma membrane proton pumps are distinguished from other pumps by various criteria (*Table 5.1*). In structure they are related to other cation-transporting ATPases.

5.2 Cation ATPases

The amino acid sequences of a number of sodium (Na^+/K^+-ATPase), calcium (Ca^{2+}-ATPase) and the plasma membrane, proton-ATPases in animal cells and K^+ pumps in bacteria have been deduced. The topography of the enzymes in the membrane, allowing channel architecture and the mode of energizing the movement of ions across the membrane are being clarified. A model showing the structure of the Ca^{2+}-ATPase of rabbit skeletal muscle sarcoplasmic reticulum is shown in *Figure 5.5*. A number of functional and structural features can be demarcated. The main feature of the 110-kd protein is a cytoplasmic headpiece. This is a prominent globular portion consisting of about 500 amino acids with a complex arrangement of α helices and β sheet structures. This headpiece contains the ATP-binding and phosphorylation domains. After ATP binding and hydrolysis, the energy released is transferred via a transduction domain to the Ca^{2+}-binding domain. The net result is the transport of Ca^{2+} across the lipid bilayer and into the lumen. The high affinity Ca^{2+}-binding domain is located around the base of the headpiece near the lipid bilayer and the amino acid sequence here determines the ion transport specificity. Binding of Ca^{2+} and concurrent phosphorylation of the enzyme leads to the transport of ions into the lumen along an uncharacterized channel. In the model shown eight transmembrane sections anchor the Ca^{2+}-ATPase firmly into the lipid bilayer and probably constitute the structure of the ion channel.

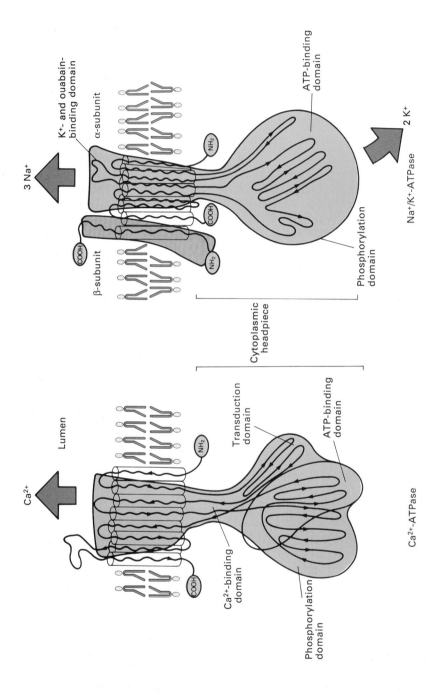

Figure 5.5. Low resolution models of the Ca^{2+} and Na^+/K^+ transport ATPases. The models emphasize the overall similarity of the α subunit of the Na^+/K^+-ATPase and the Ca^{2+}-ATPases. The knob-like shape is revealed by morphological studies.

Homologies exist in amino acid sequence at the ATP-binding, phosphorylation and transduction regions of ion-transporting ATPases. However, differences occur at the ion-binding regions and at those regions where the ions are translocated across the membrane. For example, there is about 60% amino acid identity between Ca^{2+}-ATPase and the α subunit of the Na^+/K^+-ATPases described below, thus providing molecular evidence for their evolution from a common ancestor. Even bacterial ion-transporting ATPases show some homology to the structure shown in *Figure 5.5* but the carboxyl-terminal regions are truncated.

Most mammalian cells have Na^+/K^+-ATPase pumps in their membranes. The transport is electrogenic; for each ATP molecule hydrolysed three Na^+ ions are transported out of and two K^+ ions into the cell. The α subunit is highly conserved between animals species and topographical models predict that this polypeptide may traverse the lipid bilayer eight times. In contrast the β subunit is a glycoprotein of unknown function. The β subunit traverses the lipid bilayer once, with a short 2–3-kd tail projecting into the cytoplasm (*Figure 5.5*).

The Na^+/K^+-ATPase shows specific distribution patterns on animal cell surfaces reflecting its function. In non-epithelial cells, for example fibroblasts, the enzyme is evenly distributed on the cell surface, whereas in epithelial cells the location of the enzyme on the basolateral pole underlies the vectorial transport of salts, water, organic solutes (e.g. bile acids) across the tissue.

6. Mitochondrial membranes and energy production

The membranes of the mitochondria of eukaryotic cells are specialized for production of energy, primarily ATP. As described in Chapter 1, the two membranes (inner and outer) differ in composition and function. The relative surface area of the inner mitochondrial membrane varies according to the intensity of energy production, being highly invaginated in muscle cells to form numerous cristae.

The outer mitochondrial membrane contains many distinctive proteins, some used as subcellular markers (e.g. monoamine oxidase, rotenone-insensitive NADH cytochrome c reductase) and others featuring in the importation of proteins from the cytoplasm (see Section 4.2 of Chapter 4). Porin, a 29–35-kd non-glycosylated polypeptide regulates the permeability of the outer membrane. Porin forms channels 1.7–2.5 nm in diameter that allow large molecules access to the narrow intermitochondrial membrane space.

The inner mitochondrial membrane is the site of energy transduction and is separated from the outer membrane by an intermembrane space. It surrounds a matrix where the enzymes of the Krebs tricarboxylic acid cycle, fatty acid oxidation and the pyruvate dehydrogenase complex are located. Complex assemblies of proteins, about 60 in all (*Figure 5.6*), are involved in the transport of protons and electrons, leading ultimately to the production of ATP. The properties of these membrane–protein complexes, together with the two mobile

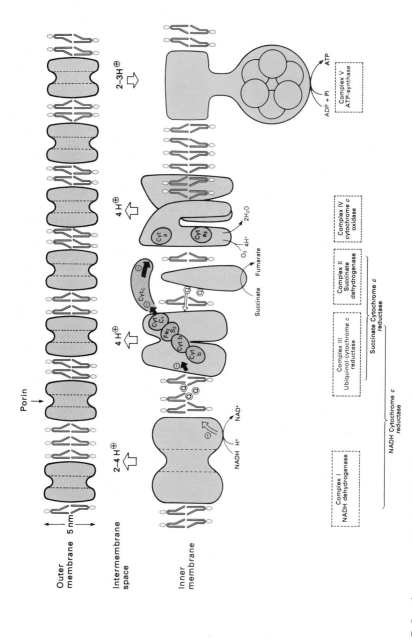

Figure 5.6. Arrangement of proteins in the inner and outer membranes of mitochondria. Complexes I, III, IV and V featuring in oxidative phosphorylation are shown (Complex II in succinate dehydrogenase). The precise structure of the various polypeptides is unclear in many cases, but the dimensions of the complexes and porin are drawn approximately to scale, with their shapes derived from low-resolution three-dimensional analyses. Q, ubiquinone. The intermembrane gap is enlarged for clarity.

electron carriers, ubiquinone and cytochrome c, are now described.

Complex I (NADH-ubiquinone reductase) contains many proteins that are poorly understood. It functions to transfer electrons and hydrogen ions to ubiquinone to form ubiquinol that then diffuses within the membrane to transfer its electrons to Complex III. Complex II (a succinate – ubiquinone reductase that oxidizes succinate to fumarate in the Krebs cycle) also contributes to the pool of ubiquinol. Complex III is ubiquinol cytochrome c-reductase. Cytochrome c functions as a mobile, relatively soluble electron carrier and is located towards the intermitochondrial space. Complex III is composed of two b cytochromes and cytochromes c and c_1, as well as a 2-iron-2-sulphur protein. The electrons transported by cytochrome c are transferred to oxygen thereby producing H_2O on Complex IV (cytochrome c oxidase, the terminal oxidase). As many as 13 polypeptides comprise Complex IV, including cytochromes a and a_3. Most studies have utilized beef heart and these show, intriguingly, that three of the subunits are coded for by mitochondrial DNA whereas the others are coded for by nuclear DNA. The passage of electrons through Complexes I, III and IV is coupled to the pumping of protons from the matrix to the intermembrane space thereby creating an electrochemical gradient comprising a pH gradient and a membrane potential. It is currently accepted, as first advocated by Mitchell, that energy from electron transport is stored across the inner mitochondrial membrane such that protons have a higher electrochemical potential outside than in the matrix. The final step is catalysed by Complex V (ATP-synthase; or F_0F_1-ATPase) which couples hydrogen-ion translocation down its gradient. It is a large multisubunit complex with a knob-like architecture characteristic of membrane ATPases (compare with *Figure 5.6*). The complex is divided into three components—a transmembrane segment containing the mechanism for transporting the hydrogen ions back into the matrix, a stalk containing a binding site for the well-known inhibitor of oxidative phosphorylation, oligomycin, and a 9 – 10-nm diameter globular domain that contains the ATP-synthase enzymes.

The properties of the energy-transducing membrane protein complexes shown in *Figure 5.6* were deduced from a variety of morphological and biochemical studies on fragmented mitochondria and on the separated components. Whilst far from being understood in structure and the mechanism of their assembly as oligomeric structures and especially in the way they interact, a common theme emerging is the complexity of their arrangement in the membrane. For example, it is thought likely that the F_0 component of the ATP-synthase from bacteria may have 28 transmembrane segments.

The electrochemical gradient is also used to pump pyruvate and $H_2PO_4^-$ into the matrix. In addition Ca^{2+} enters the matrix down a voltage gradient where it may regulate enzymes, for example the dehydrogenases.

7. Photosynthetic membrane proteins

The biochemical fixation of solar energy by plants and some bacteria is carried out by membrane proteins. In green plants, photosynthesis is carried out in specialized organelles called chloroplasts located in leaf cells. Their internal membrane system, the thylakoid membrane, contains all the components necessary for light interception and energy conversion. Thylakoid membranes are highly folded allowing a large area of membrane to be compressed into a small volume (*Figure 5.7*).

The primary process in bacterial and plant photosynthesis is a light-induced charge separation resulting in the movement, in opposite directions, of electrons and protons across the thylakoid membrane. In plants two photosystems co-operate to shuttle electrons across the chloroplast membrane. As shown in *Figure 5.7*, an oxygen-evolving complex located on the external side of photosystem II (PS II) withdraws electrons from water, thereby reducing it to molecular oxygen and protons. PS II is a multimolecular protein – pigment complex consisting of three closely associated units, with the cores embedded in the thylakoid membrane. These water-soluble proteins (OEC proteins), coded by nuclear genes, are bound to the lumenal side of the thylakoid membrane, and they play a key role in the release of oxygen. The light-harvesting chlorophyll protein complex (LHC) channels light energy to the PS II reaction centre. The core contains at least six polypeptides all encoded by the chloroplast genome, as well as several protein-bound pigments, lipids, electron carriers (quinones) and iron.

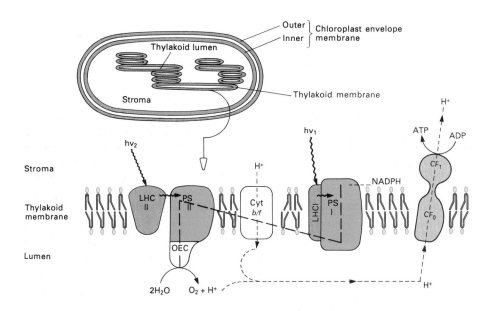

Figure 5.7. Arrangement of the five major protein complexes in chloroplast of green plants. LHCI and II, light-harvesting complexes of photosystems I and II; Cyt *b/f*, cytochrome *b/f*; CF base and CF, headpiece of ATPase coupling factor. Light energy (hν) is absorbed by light harvesting complexes and transmitted to the photosystems. The net result of the electron transport (– – – –) and proton transport (---) is ATP synthesis. OEC (oxygen evolving complex) produces oxygen from water.

The PS II complex raises the electrons to a more reduced redox potential state, leading to a cascade of redox reactions involving cytochromes b and f. This results in the conversion of a part of the gained potential energy into an electrochemical potential gradient by pumping more protons into a lumen surrounded by thylakoid membranes. As in mitochondria, the energy stored in the proton gradients is used by a membrane-bound ATPase (see below) to generate ATP. Adding to these complex reactions occurring in the thylakoid membrane is a further photosystem (PS I) which raises electrons to an energy level sufficient to generate NADPH. The chemical potential of NADPH reduces CO_2 leading to the synthesis of carbohydrates by a series of soluble enzymes located mainly in the chloroplast stroma.

The chloroplast coupling factor ($CF_0 - CF_1$ complex, see *Figure 5.7*) is an ATPase much studied by freeze-etch electron microscopy and by specific antibodies. It consists of an 11-nm head group connected by a thin stalk to a base piece (CF_0) embedded in the membrane. It is constructed of a series of three or four different subunits arranged to form a ring-like structure.

The mechanisms operational in green and purple photosynthetic bacteria are simpler than in plants, for only one photosystem acts. A cyclic electron flow in the bacterial membrane creates a proton gradient which generates ATP. Since no electrons are withdrawn from H_2O in these membranes, owing to the lower chemical potentials achieved, no oxygen is evolved.

8. Transport across bacterial membranes

Bacteria possess elaborate and varied systems for translocating essential substrates across the membrane in exchange for H^+ ions. One of the best studied bacterial membrane transporters is that in *Escherichia coli*, namely *lac* permease, that catalyses the coupled translocation of one β-galactoside molecule for one proton (a H^+/substrate symport). Thus, when an H^+ electrochemical gradient (i.e. interior is negative and/or alkaline) is generated across the bacterial membrane, the permease utilizes free energy released by the downhill translocation of H^+ to accumulate inside the cell the substrate β-galactoside against a concentration gradient. Conversely, when a concentration of β-galactoside is created in the absence of a H^+ gradient, the permease utilizes the free energy released by the downhill translocation of substrate to drive H^+ ions uphill. Thus, this mechanism is similar to that operational in mitochondria in which free energy is stored as an electrochemical gradient that can be transduced into other forms of energy.

The permease in *E.coli* has been sequenced following the cloning of the gene. It is a single and extremely hydrophobic polypeptide that traverses the bacterial membrane twelve times.

9. Further reading

Ion pumps—general

The Ion Pumps, Structure, Function and Regulation. (1988) In Stein,W.D. (ed.), *Proc. Clin. Biol. Res.* Vol. 273, A.R.Liss, New York.

Proton pumps

Bowman,B.J. and Bowman,E.J. (1986) *J.Membrane Biol.*, **94**, 83–97.
Al-Awquati,Q. (1986) *Annu. Rev. Cell Biol.*, **2**, 179–199.
Nelson,N. (1988) *Plant Physiol.*, **86**, 1–3.

Calcium channels and pumps

Meldolisi,J. and Pozzan,T. (1987) *Exp. Cell Res.*, **171**, 271–283.
Carafoli,E. (1987) *Annu. Rev. Biochem.*, **56**, 395–433.

Gap junctions

Evans,W.H. (1988) *Bio-essays*, **8**, 3–6.
Warner,A.E. (1988) *J. Cell Sci.*, **89**, 1–7.

Mitochondrial membranes

Costello,M.J. and Frey,T.G. (1987) In *Electron Microscopy of Proteins.* Harris,J.R. and Horne,R.W. (eds), Academic Press, London, Vol. 6, pp. 378–443.
Capaldi,R.A. (1988) *Trends Biochem. Sci.*, **13**, 144–148.

Thylakoid membrane proteins

Cramer,W.A., Widger,W.R., Hermann,R.G. and Trebst,A. (1985) *Trends Biochem. Sci.*, **10**, 125–129.
Rochaix,J.D. and Erickson,J. (1988) *Trends Biochem. Sci.*, **13**, 56–59.
Kuhlbrandt,W. (1987) In *Electron Microscopy of Proteins 6.* Harris,J.R. and Horne,R.W. (eds), Academic Press, London, pp. 156–207.

Bacterial membranes

Baker,K., Mackman,N. and Holland,I.B. (1987) *Prog. Biophys. Mol. Biol.*, **49**, 89–115.
Kaback,H.R. (1987) *Biochemistry*, **26**, 2071–2076.

Glossary

Affinity: the strength of binding between an antibody and an antigenic determinant, or a ligand and a receptor, or an enzyme and its substrate.

Agonist: a ligand which, on binding to a receptor, induces or stabilizes an activated conformation resulting in characteristic changes, e.g. increased ion channel conductance, enzyme activity, activation of other proteins (e.g. G-proteins).

Amphipathic: amphipathic molecules contain hydrophobic groups or sequences of groups which are spatially distinct from other groups or sequences of groups which are hydrophilic. Such molecules adopt a specific orientation in a biological membrane.

Antagonist: a ligand which, on binding to a receptor, stabilizes the ground state of the receptor.

CAM: cell adhesion molecules characterized from various tissues, e.g. liver (L-CAM) or brain (N-CAM) involved in controlling homotypic or heterotypic cell interactions.

cDNA: the DNA copy of a mRNA molecule produced *in vitro* by enzymatic synthesis and a common starting point for cloning genes from higher organisms.

Domain: conceptual division of a protein or a membrane into specific regions, not necessarily linear, with characteristic properties. Frequently positional (as in a receptor) or functional (as on the plasma membrane).

Edman degradation: a method, now automated, for analysing the amino-terminal amino acid sequence of a protein or peptide involving cyclical stepwise removal and identification of amino acids without destroying the remaining sequence.

Epitope: a determinant or an antigenic site, frequently a defined part of the linear or a spatially close amino acid sequence, to which antibodies bind.

Hydropathy plots: a term that embraces the contrary tendencies of hydrophilicity and hydrophobicity. An amino acid sequence is placed along the x-axis and the relative hydrophobicity of each amino acid is plotted along the y-axis. The plot helps determine the sequences that traverse the lipid bilayer.

Lectin: a glycoprotein of plant, animal or microbial origin which binds specifically to cell surface carbohydrate groups causing, for example, agglutination of erythrocytes or activation of lymphocytes.

Ligand: in a membrane context, a molecule or ion that binds in a highly specific manner to a receptor—includes hormones, neurotransmitters, drugs, metabolites etc. that may elicit a biological response or other effects (e.g. adhesion).

Membrane fluidity: the physical state of the bilayer, governed mainly by its lipid composition, that allows proteins and lipids to diffuse in the membrane.

Membrane turnover: the constant breakdown of proteins and lipids and their resynthesis from simple precursors. Using radioactive amino acid precursors, the half-life of a membrane protein, i.e. the time in which half the number of protein molecules are degraded, can be calculated.

Oligomeric structures: one or many proteins either identical or non-identical that are non-covalently associated to form a functional unit.

Retrovirus: a group of single-stranded RNA viruses, usually enveloped, that enter host cells via specific membrane receptors and then utilize a reverse transcriptase to synthesize a DNA that is inserted into host DNA. Transcription of the new DNA then occurs.

Sarcoplasmic reticulum: the endoplasmic reticulum of muscle cells.

Shadow replica (of cell surfaces): obtained by spreading cells out on glass coverslips, removing cell content by squirting with buffer, freeze-drying the adherent membranes and then coating them with heavy metal followed by a layer of carbon.

Signal peptide: a mainly hydrophobic amino acid sequence that helps to embed or anchor the newly synthesized protein in the membrane. It is ordinarily located at the amino-terminus and is cleaved by a peptidase after insertion of the protein into the membrane.

Site-specific mutagenesis: a method for inducing a mutagenic change at a site in a specific DNA corresponding to a known protein.

Superfamily: a diverse collection of membrane components, e.g. receptors, that share characteristic features (e.g. amino acid sequence, conformation, position of disulphide bridges) suggesting an origin from a common ancestral gene.

Topogenic sequences: amino acid sequences that direct proteins to specific locations in the cell.

Transition temperature: temperature at which membrane lipids change from a fluid to a crystalline state. It can be measured by physical techniques such as calorimetry, electron spin resonance and nuclear magnetic resonance.

Index